TOOLS OF THE ASTRONOMER

THE HARVARD BOOKS ON ASTRONOMY

Edited by HARLOW SHAPLEY *and* CECILIA PAYNE-GAPOSCHKIN

ATOMS, STARS, AND NEBULAE
Leo Goldberg and Lawrence H. Aller

OUR SUN
Donald H. Menzel

THE STORY OF VARIABLE STARS
Leon Campbell and Luigi Jacchia

GALAXIES
Harlow Shapley

EARTH, MOON, AND PLANETS
Fred L. Whipple

BETWEEN THE PLANETS
Fletcher G. Watson

STARS IN THE MAKING
Cecilia Payne-Gaposchkin

THE MILKY WAY
Bart J. Bok and Priscilla F. Bok

G. R. Miczaika and William M. Sinton

TOOLS OF THE
ASTRONOMER

HARVARD UNIVERSITY PRESS

Cambridge, Massachusetts

1961

Distributed in Great Britain by

Oxford University Press, London

Library of Congress Catalog Card Number 60-13299
Printed in the United States of America

This book was photocomposed by Graphic Services, Inc., York, Pennsylvania; printed by the Murray Printing Company, Westford, Massachusetts; and bound by the Stanhope Bindery, Boston, Massachusetts.

It was designed by Burton L. Stratton and Marcia Lambrecht Tate, members of the Harvard University Press production staff.

Preface

The information contained in most of the Harvard Books on Astronomy is based on observations of the heavenly bodies or theories relating to them. It is the aim of this volume to provide a guide to the instruments employed in the astronomical researches surveyed in the other volumes.

The achievements of modern technology have led to very rapid developments in astronomical instrumentation. Astronomy and astrophysics owe much of their recent progress to these new techniques. Frequently the advances have come from the application of methods worked out for very different purposes, such as those used in the wide field of electronics.

Scientific instrumentation today uses the services of a great variety of scientists and engineers. Only twenty-five years ago, the instruments being used in astronomical observation could have been exhaustively described in a volume of small size. This is no longer possible. The complexity and variety of modern astronomical instruments, especially those used for solar research and work in radio astronomy, impose restrictions on the treatment that can be attempted in a book of the present size.

Limitations of space have restricted the contents of this book to those techniques and instruments that are finding extensive applications in modern astronomy. The authors have tried to provide uniform coverage and have not given undue attention to their own particular interests. They have described the visual refractor, the bifilar micrometer, and the meridian circle, instruments that have had a long history in astronomical observations, as well as the equipment of the modern solar observatory and of radio astronomy. An introductory chapter furnishes a foundation for what follows for the benefit of the reader who has only a cursory knowledge of the fundamentals of optics.

There are many promising developments on the way, and some of them will ultimately become powerful tools for the astronomer. Image tubes and light amplifiers of various kinds have a great future, but application of these devices to astronomical problems is in its initial stages, and the authors decided to omit them from the book, though eventually they will without doubt become of the utmost importance. Rocket-borne and satellite-borne instruments have not been treated here. Both rockets and satellites can carry amplifiers, image tubes, and other equipment above the dense layers of the terrestrial atmosphere, and will thus overcome the absorption and turbulence so detrimental to observations made from the earth's surface. Important results have already been obtained from rocket- and balloon-borne instruments, mainly in solar problems. Larger rockets are becoming available. Satellites are already securing observations, and space platforms capable of carrying sizeable observing facilities will eventually follow. These vehicles will revolutionize the design of astronomical instruments, but to discuss this rapidly opening field is beyond the scope of this book.

The authors are indebted to many colleagues, institutes, and commercial firms for contributing photographs and diagrams.

G. R. M.
W. M. S.

Contents

Tools of the Astronomer

1

The Nature of Light

The tools of the astronomer—telescopes, spectrographs, cameras, and similar observing apparatus—are designed for the purpose of collecting, recording, and analyzing light. By the term "light" we mean radiation that we can see with our eyes and also radiation that is beyond the normal range of visibility and is detected by special recording apparatus, such as a photographic plate.

In this chapter we shall discuss the properties of light that are useful to the study of astronomy. An outline of the types of astronomical observations that can be made with the presently known properties of light will guide us in our attempt to obtain all information about the radiation from a given astronomical object. Not all of the observable properties of starlight have been employed, but most of them have been well exploited.

An outline may also suggest better instruments than those we now have. For example, television techniques and satellite vehicles are finding some application and doubtless will have great application in the future.

Light as a Wave Motion

One of the basic properties of light is that it exhibits a wave motion. With most waves, as in water, the particles of the medium in which a wave is propagated move to and fro with a constant frequency as the wave passes them. In the case of light, the motion can be regarded as being perpendicular to the direction in which the light wave is being propagated. For the present we shall speak as though actual particles were being displaced in a light wave.

Figure 1 shows examples of particles affected by a passing wave. At the instant of time for which the figure is drawn, particle A was moving upward while particle B was moving downward. Particle C was stationary. If another snapshot is taken at a slightly later time, it will be found then that the particles have moved to new positions which lie on the dashed curve at the heads of the arrows. The wave is moving from left to right. This is evident because the dashed curve is equivalent to the solid curve after this latter curve is shifted slightly to the right. The particles, however, have not moved from left to right but only perpendicular to the direction of propagation. It will be found after the interval that particles D, E, and F have the same relative positions and motions, respectively, as A, B, and C formerly had.

The foregoing description of wave motion applies to waves in a string or on the surface of water, for example. In both these cases the particles of the medium in which the wave is propagated move to and fro at right angles to the direction of displacement of the particles from their rest positions. For instance, as particle A in Fig. 1

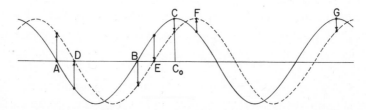

Fig. 1.   Particles of a medium are displaced by a passing wave. As the wave progresses to the dashed position the particles are displaced in a vertical direction as shown by the arrows.

moves to and fro, its displacement first increases to a maximum value, then decreases to zero, increases to a maximum in the other direction, returns to zero, and so on. A light wave can be described in much the same way if instead of a varying displacement one considers varying electric and magnetic fields.

An electric field is a region in which an electric charge experiences a force, or in which a charge *would* experience a force if it were there. Electric fields are very common phenomena even though we are usually not aware of them. For instance, there is an electric field between the wires of a power line; radio tubes and television picture tubes contain electric fields; in fact, the earth itself is surrounded by an electric field.

The earth is also surrounded by a magnetic field and so is any magnet. So also is a wire carrying an electric current. A magnetic field is a region in which forces are or would be exerted on a magnet.

The tower of a radio broadcasting transmitter is surrounded by both electric and magnetic fields, and these fields are both changing in intensity just as the displacement of a particle on the surface of a pond changes when the water is undergoing wave motion. The combination of these periodically varying electric and magnetic fields constitutes an electromagnetic wave, and it is by means of these waves that radio broadcasting and television operate. Light also consists of an electromagnetic wave motion, the only difference being, as we shall see, in wavelength.

In radiation the magnetic field is always perpendicular to the electric field. If the wave in Fig. 1 represents the electric field, the magnetic field will be perpendicular to the paper. Since the magnetic field can always be found unambiguously from the electric field, in any particular problem we need to discuss only the electric field. Hereafter in this book, when we talk about the direction of the vibration of light wave we shall be referring to the direction of the electric field.

To summarize, a light wave being propagated through a vacuum, air, glass, or most other transparent materials, consists of electric and magnetic fields which are perpendicular to one another and also to the direction in which the wave is moving. These fields are constantly changing their intensity. An electrically charged particle, if placed in the path of the wave, would vibrate with the passing of the wave just as the water at the surface of a pond does with the passing of a water wave.

An important property of a wave is its velocity or rate of travel. A definite point on a wave, a crest for example, moves along at a definite rate. In Fig. 1 the crest of the wave is at the point C, but by the time t later it has moved to the point F. The velocity of the wave is the distance CF divided by the time interval t. For light waves moving in a vacuum this velocity has always the same value, irrespective of the distance between crests.

The discovery that light consists of electromagnetic waves came about from a comparison of the measured velocity of light with the velocity of electromagnetic waves predicted from theory by the physicist James Clerk Maxwell. Maxwell's theory predicts a velocity of 299,860 kilometers per second for electromagnetic waves.

The velocity of light has now been very accurately measured in the laboratory. One of the methods that has been used is to interrupt a beam of light by a rotating wheel that has a number of teeth evenly spaced about its rim. The light beam is then made to traverse an accurately measured distance and is reflected back to the wheel by means of a mirror. The speed of the wheel may be adjusted so that a tooth in the wheel will come into position to block the returning beam in exactly the time it takes the light beam to traverse the measured path and return to the wheel. The path difference divided by the time it takes the wheel to advance from a gap to a tooth gives the velocity of the light beam. A recent refinement of this method, using an electronic light interrupter in place of the wheel, gave 299,792.7 ± 0.3 kilometers per second for the velocity of light. This velocity is close to the value predicted from electromagnetic theory, which, however, is only approximate because of the difficulty in accurately measuring a certain electrical constant.

A better agreement is obtained by comparison of the measured velocity of light with the velocity of radio waves, which are known, from the method by which they are generated, to be electromagnetic waves. A recent determination of the velocity of radio waves across a desert flat yielded a value of 299,795.1 ± 3.1 kilometers per second. This velocity does not differ by more than its probable error from the velocity obtained for light waves. Because it is definitely established that light is an electromagnetic wave, we will use the term light to mean electromagnetic waves whether or not the radiation can be seen.

The distance between two points, such as C and G in Fig. 1, which are on identical parts of two adjacent waves, is an important characteristic. This distance, called the wavelength, is designated by the Greek letter λ (lambda). The wavelength is the property which indicates whether the particular light is in the visible region or in another part of the electromagnetic spectrum. The wavelengths of visible light lie between 4×10^{-5} and 7.5×10^{-5} centimeters—a range of less than one octave. On the other hand, radio waves extend from wavelengths of a few millimeters to 20,000 meters and longer. The interval between these two regions, that is, from 7.5×10^{-5} cm to a few millimeters, is termed the infrared portion of the electromagnetic spectrum. Wavelengths just short of the visible range are in what is termed the ultraviolet region. A unit of length that is more suitable than the centimeter for the visible and ultraviolet regions is the angstrom unit, abbreviated A, and equal to 10^{-8} cm. Thus the visual spectrum extends from 4000 A, to 7500 A, the ultraviolet from about 100 A to 4000 A.

There is a relation between the wavelength of a wave, the velocity of the wave, and its frequency ν (the number of vibrations per second). The velocity of light is the product of the distance traversed in one vibration by the number of vibrations per second; hence $c = \lambda\nu$. Because the velocity of light in vacuum has always the same value, a particular wavelength is always associated with the same frequency. Radio waves of 10 centimeters wavelength execute 3×10^9 vibrations per second; a light wave of 3000 A has a corresponding frequency equal to 10^{15}. The frequency of 3×10^9 vibrations per second is generally referred to as 3000 megacycles per second. The megacycle, abbreviated Mc, is a million cycles.

The wavelength measured by an observer is not immutable, but depends on the relative velocity between the source and the observer. Suppose that the source that has emitted the wave of Fig. 1, which is assumed to be travelling to the right, is approaching from the left with a velocity v. If the source is at the crest at the extreme left of the figure, it will have moved a distance v/ν to the right since it emitted the crest of the wave now at C. Thus the wavelength is v/ν shorter than it would be if the source was stationary. Using the relation $\nu = c/\lambda$, we find that the new wavelength is $\lambda' = \lambda (1 - v/c)$. This expression is accurate only when v is less than 0.1 c, at which point it gives a 5-percent error in the change. Hence the light, because the wavelength is shorter, appears to be bluer when the source is approaching the observer. This relation also applies when the observer is approaching the source, and in fact, it applies whenever the velocities of the source and observer along the line between them are different by v. But if the source is moving away from the observer, then the minus sign in the relation becomes a plus sign, and the light appears redder to the observer. Since most velocities encountered in practice are much less than the velocity of light, the wavelength changes are very small. But these changes, which are named for the Austrian physicist J. C. Doppler, are very important to the study of astrophysics, for they reveal the velocities of celestial bodies.

Two separate trains of waves having exactly the same frequency and traversing the same space may under the proper conditions produce the phenomenon known as interference. If the two trains of waves have the same amplitude (that is, if the distance C_oC in Fig. 1 is the same for both), and if their electric fields at any given instant are in opposite directions, the two waves will cancel each other and the effect will be the same as if there were no waves at all. Thus, if two trains of light waves cancel each other in this way the effect is darkness. This phenomenon is called destructive interference. If, on the other hand, the electric fields are in the same direction, the two wave trains will reinforce each other and the effect produced will be greater than that produced by either one alone. This is the phenomenon of constructive interference. Waves that interfere constructively are said to be in phase with each other. If two waves are not in phase with each other, the phase difference between them is expressed by the distance, measured in wavelengths, from a point, say a maximum of one wave, to the corresponding point of the other wave. For example, if in Fig. 1 the two curves are taken to represent two waves, the phase difference between them is represented by the distance AD expressed as a fraction of the wavelength. If the phase difference is zero or a whole number of wavelengths, the waves are in phase and constructive interference results. Destructive interference occurs when the waves are out of phase by $\frac{1}{2}\lambda$, $\frac{3}{2}\lambda$, $\frac{5}{2}\lambda$, or any other odd multiple of a half wavelength. Waves that are out of phase by less than $\frac{1}{2}\lambda$ also interfere and the effect produced is intermediate between complete cancellation and constructive interference.

It was mentioned in the preceding paragraph that two trains of waves must have exactly the same frequency if they are to interfere. Waves from the same source do have the same frequency and thus can be made to interfere if they can be brought together in the same space. Two ways of accomplishing this splitting and recombination for light are illustrated in Fig. 2. In the Lloyd's mirror experiment, illustrated in Fig. 2(a), part of a beam of light is reflected from a mirror in such a way as to fall in the same area that is traversed by the undeviated part of the beam. In this diagram, solid lines represent the crests and broken lines the troughs of the waves. According to the previous discussion, where two crests coincide there is constructive interference of the waves and maximum electric field occurs. At points where a trough and a crest coincide the two waves cancel and so destructive interference occurs. Where two troughs coincide there is again constructive interference and so maximum electric field is again obtained but this time in a negative direction.

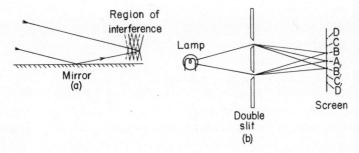

Fig. 2.   Interference of light: (a) Lloyd's mirror; (b) Young's experiment.

The other method of producing interference, shown in Fig. 2(b), is Young's experiment. A source of light, at the left, illuminates two narrow parallel slits. Since these slits are illuminated by the same source, they behave like two sources between which there is a constant phase difference (which may be zero, though this is not necessary). Now on account of diffraction (see a later section of this chapter), light is emitted from these slits in all forward directions. Let us now consider the illumination of the screen on the right by the two sources. At a point A at the center of the screen the distances from the two sources are identical. The waves arriving here will be in phase and there will be a maximum of light. At points B and B' the distances from the two slits differ by a half wavelength; the two waves destructively interfere and there will be no light. At points C and C', where the distances differ by one wavelength, the waves constructively interfere again and there is light. At D and D' the two waves are again out of phase and there is no light. Thus on the screen one sees a series of light and dark bands, called interference fringes.

The Electromagnetic Spectrum

We have already mentioned the fact that visible light and radio waves are both forms of electromagnetic waves. One proof that the two are the same kind of wave is the equivalence of their velocities. These various waves have a wide range of wavelengths. A diagram of the experimentally observed electromagnetic spectrum is shown in Fig. 3. The spectrum extends from gamma radiation (which comes from the nucleus of an atom) through x-rays, ultraviolet rays, visible light, and infrared radiation to radio waves. The wavelengths that have been observed extend from less than 10^{-11} cm in the region of gamma rays up to 20 kilometers or longer in the radio region. The entire range between these wavelengths has been experimentally produced in the laboratory.

We are familiar with the fact that a very hot body emits visible light. For example, a piece of iron can be heated to white heat by a torch. If it is allowed to cool, it turns yellow and then red and finally becomes dark although it may still be hot. But it is dark only because our eyes do not see the infrared radiation that it is emitting. If a body is heated, the intensity of light emitted in the infrared increases as the temperature rises but the visible radiation emitted increases much more rapidly. The object becomes first red, then yellow, and finally white because the blue part of the visible spectrum increases much more rapidly with a rise in temperature than does the red part of the spectrum.

A piece of iron emits radiation according to well-established laws. The intensities of the emitted radiation are distributed smoothly through a wide region of the spectrum. Radiation of this sort is said

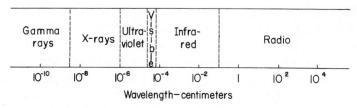

Fig. 3.   The electromagnetic spectrum. Astronomical observations have been made in the spectrum from the ultraviolet portion to the radio portion.

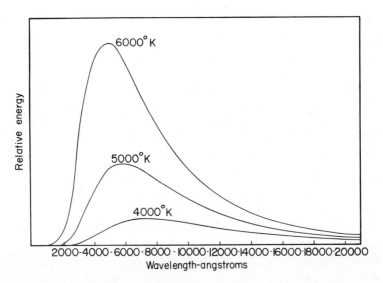

Fig. 4. The continuous spectrum emitted by hot black bodies.

to be continuous radiation. The continuous radiation that is emitted by a special kind of heated body, called a black body, is of great fundamental importance. A black body is defined as any body that absorbs all radiation incident upon it regardless of wavelength, without reflecting or transmitting any of the incident radiation. A black body may not appear black to the eyes (using the dictionary definition of the word "black") because it may be hot enough to appear red or yellow or even white.

The total amount of radiation emitted per square centimeter of a black body depends on the temperature; it is not distributed equally throughout the spectrum. Figure 4 shows graphically how both the distribution and the total amount depend on the temperature. At a given temperature the radiation reaches a maximum value and declines gradually toward zero on either side. The wavelength of the maximum depends only on the temperature and may be found from the simple formula $\lambda_m = 0.2880/T$ cm, where T is the absolute temperature (found by adding 273 to the temperature in degrees centigrade; the temperature of ice is 273° absolute and that of boiling water is about 373° absolute). The shift of the maximum of the curve toward shorter wavelengths as the temperature is increased is responsible for the light from heated objects turning from red to orange to yellow to white as they are heated. Before an object becomes red hot, it is infrared hot and emits copiously in the infrared. Even cold objects emit radiation in the radio range of the spectrum. The sun is approximately a black body but at certain wavelengths strong absorption of the radiation occurs in the cooler solar atmosphere, causing marked departures from black-body conditions. The thermal radiation from the sun has been observed not only in the visible and infrared but even in the radio spectrum at wavelengths near 1 centimeter and 1.5 mm.

Radiation from hot gases has generally very different characteristics from the continuous emission described above. If the gas is in a state in which only atoms are present and if the temperature is high enough and the density low enough, then light is emitted only at certain very narrow ranges of wavelengths which are characteristic of the elements present in the gas. For example, in Fig. 5(a) is shown the spectrum produced by atoms of vaporized iron. Radiation occurs only in narrow, sharply defined regions, called "lines" from their appearance; they are images of the slit of a spectrograph used to study the characteristics of the light (see Chapter 6).

Molecules also emit radiation in narrow ranges of wavelength, but the resulting lines in the spectrum occur in groups, called bands. Figure 5(b) shows band emission by the molecule NO. Such simple molecules emit the lines of a band in a regular progression, as the figure demonstrates. However, the lines are sometimes so close together that they are not resolved—separated by as much as their own width. They then have the appearance of Fig. 5(c), where the individual lines are not seen but only the cumulative effect of each band.

Light Quanta

Until about 1900 the theory that light consists of electromagnetic waves appeared to explain all optical phenomena with few exceptions. One of these exceptions, however, was the energy distribution in the black-body spectrum. The theory of that time predicted a distribution of energy with respect to wavelength that was clearly different from the observed distribution given in Fig. 4. This and the few other exceptions to the electromagnetic theory have served to demonstrate that light is a considerably more complicated phenomenon than had been supposed at the turn of the century. This complication has been fortunate for astrophysics. Many of the recent discoveries have come about from the new understanding of light.

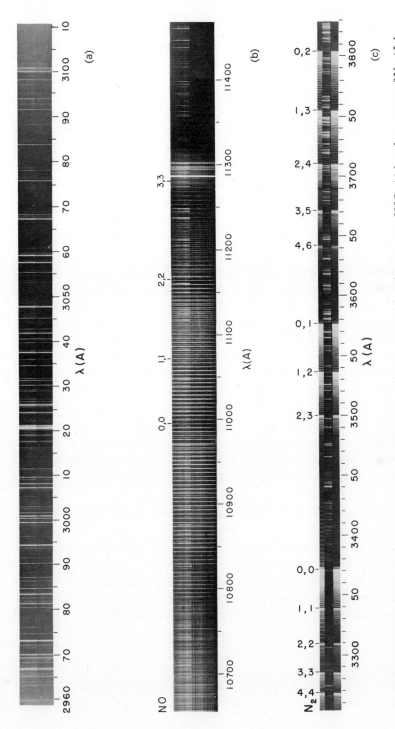

Fig. 5. Different types of laboratory spectra: (a) line spectrum of iron vapor; (b) band spectrum of NO; (c) band spectrum of N_2. (Johns Hopkins University.)

In the year 1900 Max Planck found that the spectral distribution of radiation shown by Fig. 4 could be explained only if light consisted not of steady trains of waves but of bursts of energy, each of amount hv, where v is the frequency of the light and h is a constant which is now called Planck's constant. These bundles of energy are called *photons* or *quanta,* and they have some of the characteristics of particles rather than waves.

Another phenomenon that helped to promote this change of view regarding the nature of light was the photoelectric effect. It was discovered that light falling on certain metals releases electrons from the surface of the metal, and that the energy of these electrons does not depend on the intensity of the illuminating light but only on its frequency. Moreover, even under very weak illumination, where the energy is provided very slowly, an electron may be emitted with high energy (if the frequency is high) almost immediately after the illumination is turned on, long before the energy that it carries could be absorbed from a train of waves. The explanation of the paradox is that the light comes in quanta or photons and such a particle may be emitted immediately after the light source is turned on. The average rate at which the photons are received by the metal, however, is that required by the weak light intensity. The energy of the emitted electron is given by the expression $E = hv - W$, where E is the energy of the emitted electron and W is a constant that depends on the material from which the electron is emitted, it is essentially the energy lost by the electron in the process of getting through the surface of the metal. This expression was published by Einstein in 1905 and it was for this discovery that he received the Nobel Prize in 1921.

But we now have another paradox, since we have two different theories of light, one which states that light consists of electromagnetic waves, a theory that explains a large number of phenomena, and the other, the photon theory, which states that light consists of photons. These photons are required to explain some of the phenomena of light but cannot explain all. For all of the other phenomena which cannot be explained by the photon theory, the electromagnetic theory is required. Are the two theories mutually exclusive? We note that the photon theory used the concept of frequency, a concept coming from the wave theory, to obtain the energy of the photons. But how can light be viewed as both particles and waves? This question has led to the present view that light comes in wave

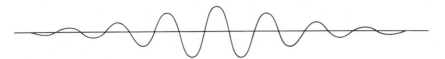

Fig. 6. A wave packet—the reconciliation of the electromagnetic theory and the quantum theory of light.

packets. Figure 6 shows schematically such a wave packet. It is a wave that is terminated on both ends so as to be like a particle. Its length and amplitudes are such that the total energy contained in the wave is equal to $h\nu$. Laws of the newly developed mechanics which resulted from this discovery about light show that a photon cannot be absorbed in part. Either all of the photon is absorbed or none is absorbed.

In the remainder of this book we shall sometimes find it convenient to discuss light as waves, when interference is concerned, and at other times to consider light to be particles, especially when we are discussing the photoelectric effect in Chapter 5. How these two theories dovetail together will be demonstrated by examining the phenomenon of polarization of light. In order to explain polarized light and unpolarized light we shall find that both theories are necessary.

The wave depicted in Fig. 1 lies entirely in the plane of the paper. Since its electric field lies in one plane, it is said to be plane polarized. A beam of light in which the electric fields of all the wave packets, or photons, lie in a single plane is plane polarized, and the plane that contains the electric field is called the plane of polarization. Although a plane-polarized wave can be easily produced, it is not the kind of light wave that occurs most frequently. *Natural* light (that produced by an incandescent bulb or the sun, for example) is not plane polarized but is unpolarized. In this case the planes of polarization of the photons are oriented at random, so that no one direction is commoner than any other.

One of the most convenient optical devices for producing or testing for polarized light is the material called by the trademark name Polaroid. Polaroid has the property of transmitting 80 percent of the light whose plane of polarization is in one direction and of absorbing almost all of the light that is polarized in the direction perpendicular to this. For polarizations between these angles, only part of the light is transmitted. Figure 7 shows the action of Polaroid.

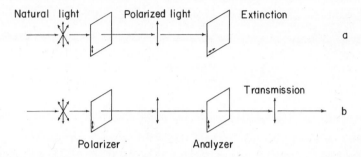

Fig. 7.   Polarization of light by sheet polarizers: (*a*) polarizers oriented at 90° to each other transmit no light; (*b*) polarizers parallel to each other transmit light.

Natural light enters from the left and falls on a piece of Polaroid. In this drawing the Polaroid is oriented to absorb all of the radiation polarized in a plane perpendicular to the page while it transmits most of the light polarized in the plane of the page. Thus it produces polarized light from natural light. The second Polaroid may be oriented to transmit only that light which is perpendicular to the plane of the page, in which case all of the light is stopped as in Fig. 7(*a*). The Polaroid may also be oriented again so that its direction of transmission is parallel to the page, as in Fig. 7(*b*), in which case most of the light is transmitted. The first piece of Polaroid is frequently referred to as the polarizer and the second piece is referred to as the analyzer when they are used in an arrangement like that in Fig. 7.

Circular and elliptical polarization have not yet been discussed. In the case of circular polarization the electric field, instead of varying in strength as the wave progresses, continually rotates about the direction of propagation. This situation is illustrated in Fig. 8(*a*).

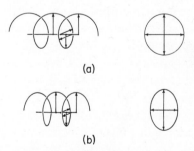

(a)

(b)

Fig. 8.   Diagram to illustrate polarization: (*a*) circular; (*b*) elliptical.

Plane-polarized light may be converted to circularly polarized light by passing it through a device called a quarter-wave plate. This frequently consists of a sheet of mica which hås been split to just the right thickness. When it is used as a quarter-wave plate the direction of the polarization of the incident light must make an angle of exactly 45° with one of the crystalline planes of the mica crystal.

Elliptical polarization is similar to circular in that the direction of the electric field rotates, and in addition the electric field varies in strength as shown in Fig. 8(b). As the field rotates, the end of an arrow representing the strength of the electric field, or the displacement of the wave, describes an elliptical helix through space. In both circular and elliptical polarization the direction of rotation may be either clockwise or counterclockwise as the wave progresses. If the direction of rotation is to the right when one looks in the direction that the wave is progressing, the wave is said to be right-hand circularly polarized, and it is said to be left-hand circularly polarized if the field rotates to the left. These terms, right-hand and left-hand, are also applied to elliptically polarized light.

Reflection, Refraction, and Diffraction of Light

Reflection. Surfaces that are shiny or glossy are said to exhibit specular reflection and are especially important to astronomical instruments. They are usually made by polishing glass smooth and then coating it with a thin metal layer which preserves the polish and adds a highly reflecting coat (see Chapter 3). Reflection of light from a polished surface follows a simple optical law. A light ray falling on the surface at the point O (Fig. 9) is reflected in such a way that the incident ray AO and the reflected ray OB make the same angle with the normal or perpendicular ON to the surface. More-

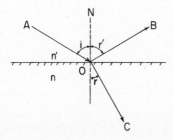

Fig. 9.   Reflection and refraction of light at an air-glass interface.

over, the two rays AO and OB and the normal ON all lie in the same plane. This law of reflection is important to the construction of telescopes that use mirrors.

Refraction. When a ray of light falls on the surface of a transparent optical medium such as glass, in general only part of the light is reflected; the rest enters the glass. Thus in Fig. 9 the part of the incident ray AO that enters the glass is bent toward the normal to the surface, and results in the refracted ray OC. The angle of refraction r may be found from the relation generally known as Snell's law:

$$n' \sin i = n \sin r,$$

where i is the angle of incidence and n' and n are numbers that are characteristic of the air and the glass, respectively, and are called the refractive indexes of the two media. For vacuum, $n' = 1$, and for air it is so close to unity that the simpler expression $\sin i = n \sin r$ can usually be used. As with reflection of light at a surface, the incident ray, the refracted ray, and the normal to the surface all lie in the same plane.

Let us assume that n' is greater than n. Then from Snell's law r must be greater than i. But if i is already rather large, then $\sin r$ may be greater than unity. If this is the case, one cannot find a value of r that will satisfy the equation, since no angle has a sine greater than unity. What happens in this case is that no light enters the second medium but all of it is reflected, the angles of incidence and reflection being equal as always. The phenomenon is called total internal reflection, and the value of i that makes $\sin r = 1$ is called the critical angle; it is defined by the equation $\sin i_c = n/n'$.

Total internal reflection is taken advantage of in the Nicol prism, a device for producing polarized light. Such a prism is constructed from two pieces of calcite crystal cemented together as in Fig. 10(a). Calcite is birefringent, that is, it has the property of breaking up an incident ray of light into two refracted rays, which are bent by different amounts on entering the crystal and are polarized in mutually perpendicular planes. When the rays reach the cemented face, the one whose plane of polarization is perpendicular to that of the figure is totally reflected at this surface so that it passes out through the side of the prism and is lost. The other ray, which is polarized parallel to the plane of the figure, continues on through the prism. The efficiency of this kind of polarizer is high; the transmission is about 92 percent for incident light polarized in the favored direction. A

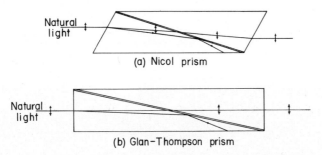

Fig. 10.   Polarizing devices: (*a*) Nicol prism; (*b*) Glan-Thompson prism.

modification of the Nicol prism is the Glan-Thompson prism, which has end faces perpendicular to the entering and emerging beam, as in Fig. 10(*b*). The Glan-Thompson prism requires more calcite for its construction, however.

The refractive index of a transparent medium depends on the wavelength of the incident light. For most materials the index is higher for violet light than it is for red light, that is, the angle of refraction is greater for violet light than for red light. This means that a beam of light containing many wavelengths will be separated after it enters the refractive medium into its component wavelengths, a phenomenon called dispersion.

Diffraction. On the last few pages we have assumed that light follows straight lines, which we have called rays. But if we try to define a certain ray by means of an aperture through which a beam of light must pass, we find that the light after passing through the hole begins to spread out, and the smaller we make the aperture, the more the light spreads. This phenomenon—the spreading of light beams as they pass through apertures—is known as diffraction.

It is not necessary to have very small holes to cause the spreading; even the aperture of a relatively large telescope objective will cause appreciable diffraction. It is not even necessary that we have an aperture in order to produce the effect. An opaque straightedge interposed in the beam will do it. Diffraction of light by a circular aperture is illustrated schematically in Fig. 11(*a*). The appearance of the pattern formed by the light when it strikes a white screen is shown in Fig. 11(*b*). Note that the light is not uniformly spread after passing the aperture, but bunches together into a pattern and is absent in some areas. The pattern is due to interferences of the light coming from different parts of the aperture.

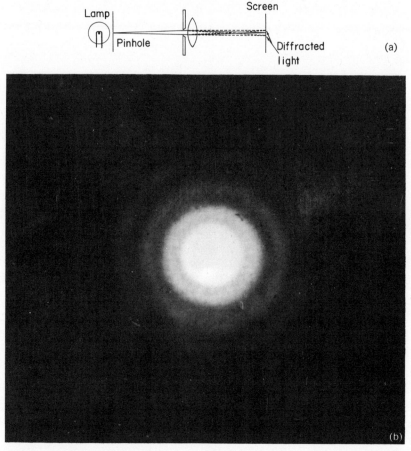

Fig. 11.   Diffraction of light by a circular aperture: (*a*) schematic diagram; (*b*) photograph of pattern of light on the screen.

The cause of diffraction is in the wave nature of light. The amount of diffraction in a given case depends on the ratio of the wavelength of the light to the size of the aperture. Diffraction increases proportionally to wavelength and inversely proportionally to the diameter of the aperture. The diameter of the lens or the mirror of a telescope defines an aperture for the light and so produces diffraction. In a later chapter we shall discuss the effects of this diffraction in telescopes. With radio telescopes, although the apertures are large, the ratio of the wavelength to the size of the aperture is a great deal larger than with visual telescopes. Consequently the diffraction is very much greater than it is with even the smallest of telescopes designed for the visible wavelengths.

The diffraction of light is used in spectroscopy in the diffraction grating. The construction of gratings and how they work are discussed in Chapter 6.

Observable Properties of Starlight

Now that we have an understanding of the nature of light we shall outline those properties that can be observed astronomically. Early astronomers were content to record only the position of an object and add a very rough estimate of its brightness. In the last 50 years, there has been an increasing emphasis on astrophysics, and the astronomer is now very much concerned with making every possible type of physical measurement on the light emitted by the object he is studying. The remaining chapters of this book are devoted to apparatus for making these measurements.

Intensity. The first physical property of starlight to be measured was the intensity of the light. The total energy emitted by a star cannot be measured because the earth's atmosphere absorbs light of many wavelengths either partly or completely and because the detectors or measuring instruments that are employed are sensitive to some wavelengths and not to others. However, much important information has been learned about distances and temperatures of the stars and the variability of their light, through measurements of the intensity in limited spectral regions. Highly precise instruments, called photometers, have now been developed for use with telescopes to measure the intensity of starlight.

Wavelength. One of the most fruitful types of measurement that has been made of astronomical objects is that of the distribution of light throughout the spectrum. We have already seen (Fig. 5) some laboratory spectra. Spectra of stars are shown in Figs. 135 and 139. One of the first uses to which stellar spectra were put was the determination of the various chemical elements present in the stars from a comparison of wavelengths of the lines observed with wavelengths of lines measured in the spectra of elements obtained in the laboratory. Not only have the spectra of stars revealed their qualitative composition, but they have also permitted determinations of the quantities of the elements present, the temperatures of the atmospheres of the stars, the presence and strength of magnetic fields of some stars, the velocities of the stars toward or away from us, and the pulsations or variations in size of stars. Nearly two-thirds of the time of the largest reflecting telescope, the 200-inch Hale reflector on Palomar

Mountain, is spent in taking spectra of the stars rather than photographing star fields or for photoelectric observing. The kinds of spectrographic apparatus that are used in conjunction with telescopes are discussed in Chapter 6.

Polarization. The light emitted by the stars is believed to be unpolarized. However, the light of some distant stars has been found to be slightly polarized by passage through the dust of our galaxy. The light from some of the planets is partially polarized. Recently it was discovered that the light from parts of the Crab Nebula is strongly polarized; in some regions the polarization is nearly 100 percent. Equipment for measuring the percentage of polarization of the light from stars is treated in Chapter 5.

Elliptical and circularly polarized light has not yet been detected in the total light or the visible light from astronomical objects, but such polarizations have been discovered in the radio region in radiation from the sun and from Jupiter. However, circular and elliptical polarizations have important occurrences in the highly dispersed spectra of the sun and of certain stars that give evidence of large superficial magnetic fields. Through the effect of the magnetic fields it is learned that some lines in stellar and sunspot spectra are broken into two components which are right- and left-handed elliptically or circularly polarized. This splitting into components is known as the Zeeman effect and is illustrated in spectra of a magnetic star in Fig. 148.

Other Properties. The foregoing discussion includes all of the properties of light that have been extensively investigated in astronomical objects. For the sake of completeness two other properties, which either have not been measured or are not regularly measured, are mentioned here. One of these properties is the velocity of light, which might be considered to depend on the source; at the present time, however, no connection between them is known to exist. The other property is the phase of the light. It is, of course, hopeless to expect to know the total number of waves between the observer and the object. However, it is conceivable that the relative phase difference between two different parts of the same object can be measured. It is believed that all of the astronomical objects that are observed have continuously and rapidly varying phase differences between any of their parts (incoherence) and so no interference of light coming from two areas can occur. Incoherence may not exist with all of the radio sources and it is worth while to keep the possibility of coherence in mind.

Pictorial Representation

A telescope objective lens or mirror such as the ones described in Chapter 3 provides an image of an astronomical object. The first observations were visual ones and were merely qualitative. The image provided by the objective was observed with the aid of an eyepiece, and for extended objects such as planets or nebulae drawings were made. These were not accurate owing to the astronomer's inability to portray correctly what he saw and because his prejudices probably entered into the observations he was attempting to make. With the advent of photography more accurate representations of the images of astronomical objects could be obtained. In addition, the relative positions of stars were accurately reproduced on photographic plates which provided a permanent record and could be measured and studied after the observations were completed.

As sources of light, astronomical objects are of two kinds—point sources and extended sources. A star does not show any appreciable disk of its own but gives a more or less pointlike image on a photographic plate. We shall see later that star images actually have a certain size; however, this size is not related to the diameter of the star itself but is of instrumental origin. Objects of the second type, such as nebulae and planets, extend over appreciable areas. The telescopes that are required to image most efficiently these two types of objects are not the same. The requirements for making the best photographs of them will be discussed in Chapter 3.

We can measure not only the intensity and position of star images on a stellar photograph, but in regions where continuous images occur we can determine the light distribution that contributes to the original image. Generally the variation in light is represented by drawing lines, called *isophotes* or *intensity contours*. These are similar to the elevation contour lines which are drawn on many maps. On an intensity-contour drawing each of the lines passes through points where a specific amount of light fell in the original photograph. The contour representation yields quantitative information from a pictorial representation. For example, one may be interested in merely the light intensity at each point of a nebula, or one may want to know the total area of a nebula between two light-intensity contours. Such quantitative information is easily obtained from a contour representation. Machines have been developed for automatically obtaining the isophotes from a photograph (Fig. 111).

Atmospheric "Windows"

Unfortunately for astronomy, the atmosphere is not transparent to all wavelengths in the spectrum. But it is transparent to enough wavelengths to give us a fairly widespread coverage of the spectrum. Figure 12 shows graphically the portions of the spectrum that are transmitted by the atmosphere; the portions that are not transmitted are indicated by crosshatching. Wavelengths shorter than about 3000 A are completely absorbed. The "window" that includes the visible range begins at this wavelength and extends up to about 20,000 A but some narrow bands in the near infrared are heavily absorbed by water vapor. There is another window of moderate importance between 8 and 13 microns (80,000 to 130,000 A) which has been used for observations of the planets. The radio window extends from about 1 cm to approximately 15 meters but the wavelength of this 15-meter cutoff is variable. The cutoff is caused by reflection of radio waves by the ionosphere, a layer of the earth's atmosphere containing ions and electrons. The opaqueness of our atmosphere to other portions of the spectrum is caused by absorption by atmospheric gases. The cutoff at wavelengths shorter than 3000 A results from absorption by ozone; but other gases, such as oxygen and nitrogen, maintain this opacity to much shorter wavelengths. The spectrum between 13 microns and 1000 microns, or 1 mm, is strongly absorbed by water vapor.

All of the above-mentioned transparent regions are now being exploited for astronomical work; the last part to succumb was the radio region. Observation in this area is now progressing rapidly. Chapter 8 discusses the special equipment that is needed for radio astronomy.

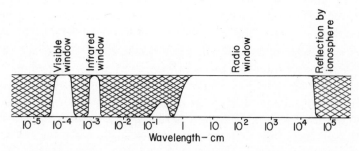

Fig. 12.   The transmission of electromagnetic radiation by the earth's atmosphere. Crosshatching indicates regions of absorption.

2

Photography

Present-day research utilizes photography in many ways. One of the oldest applications of the photographic process for scientific research took place in astronomy. It had a tremendous impact on the development of this science, opening entirely new fields, otherwise completely inaccessible. The new observational techniques called for specially designed equipment and led to great achievements in the field of optics.

Besides the human eye, the photographic plate was the only light detector used for recording the radiation of the stars until the various photoelectric cells of recent date. In March 1840, very shortly after the daguerreotype process was announced, John William Draper of New York made a daguerreotype of the moon. Ten years later, on the night of July 17, 1850, George Phillips Bond and John A. Whipple took the first photograph of a star. The telescope they used was a refractor of 15-inch aperture at the Harvard College Observatory in Cambridge, Massachusetts. The instrument was

Fig. 13.   The 15-inch refractor of the Harvard College Observatory, with which the first photograph of a star was taken in 1850.

made by Merz in Munich for visual work (Fig. 13). Bond himself realized immediately the enormous advantages of the photographic method. The plates represent a lasting record which can be studied thoroughly and repeatedly at any convenient time; a single photographic plate can contain stored information of many thousands of stars.

The photographic plate has a property which the human eye lacks. It accumulates the light that falls upon it while the plate is exposed, and as a consequence the photographic image continuously becomes stronger. Thus the photographic technique can make faint objects visible which the eye is not able to see.

The photographic process involves three major steps: the formation of the image by optical means upon a light-sensitive emulsion; the development of the latent image by chemical treatment to produce a visible one; and removal of the unexposed and unchanged photosensitive material (fixing the image).

The Emulsion

It has long been known that silver salts darken upon exposure to light. Silver salts have become the basic material for making light-sensitive emulsions which are coated on glass, film, or paper as a base. We are concerned here only with glass and film as bases.

The sensitive coat, the emulsion, is a suspension of the silver salt crystals in gelatin, the salts usually being silver bromide containing a little silver iodide. The making of an emulsion is a delicate process, requiring the utmost care to obtain the desired properties. Although uniformity is the aim, differences inevitably occur between batches of emulsion. Too many circumstances make it hard to reproduce repeatedly exactly the same properties of the emulsion. Therefore the features of an emulsion should be individually determined for each delivery of plates to be used for scientific purposes and the characteristics of the emulsion can then be properly allowed for in exposure and development.

The emulsion is coated on its carrier, glass or film. Ordinarily, the sensitive layer is about 2 to 5 microns thick, on a glass base ranging from 1 to 5 millimeters or more in thickness, depending upon its size. Thicker glass bases are sometimes used to prevent warping, if the highest accuracy in measuring the positions of celestial objects is desired. The emulsion swells by wetting during the development and shrinks in drying. The strain exerted can be sufficient to bend the film or even the glass plate. Film, and sometimes glass plates, are therefore coated on the back with gelatin. The shrinkage of this backing balances the effects of the shrinkage of the emulsion on the front of the base.

For astronomical and other scientific purposes, attention must be paid to the quality of the glass used as the base for the emulsion. Ordinarily, plates are made from good window glass. The glass should be accurately flat and free from air bubbles affecting the surface. Window glass used to be manufactured by a process producing sheets that were not perfectly plane but had a slightly concave and a slightly convex side. When used for photographic plates, the concave side was coated with the emulsion, leading to somewhat greater thickness in the middle than at the edges of the plates. Modern window glass rarely shows a concave side but often has a slightly wavy surface. Coatings of nonuniform thickness result, and are frequently not acceptable. In such a case, the glass ought to be carefully inspected and pieces selected for better flatness (mirror glass). Departures from flatness of about 0.001 centimeter per linear centimeter might be adequate for stricter requirements.

For some purposes, such as are encountered in positional work of the highest accuracy, even this flatness tolerance is too much and specially polished flat glass must be used. With the help of optical instruments for inspection, flatness of the sheets within 0.0006 cm per linear centimeter can be obtained. If glass of this degree of flatness is employed, it must be protected against deformation either under its own weight or by the strain applied in the plateholder. Otherwise, the effort of selecting it would be wasted. Adequate rigidity requires that the glass have a certain thickness. However, many modern cameras call for plates bent to the shape of the focal curve. For this purpose glass with thickness of 0.75 or 1.0 mm is used. It is inconvenient to handle because of its tendency to break, particularly if the sizes are large.

Various coatings other than gelatin are sometimes added to the emulsion. Since emulsions are quite sensitive to mechanical action, such as scratching or pressure, an over-coat of a transparent anti-abrasive layer is occasionally provided to protect them. Also, the light-sensitive coatings are sometimes made of layers of emulsions of different sensitivity, in order to extend the useful exposure range of the emulsion.

Since an emulsion does not adhere well to a cellulose film, a substratum is coated on the film base before the emulsion is put on. The substratum contains an adhesive and a solvent. It quite frequently changes the characteristics of an emulsion, which is the reason why the same emulsion coated on glass displays slightly different properties than when coated on film. Occasionally the glass

Fig. 14.   Halo around the photographic image of a bright star.

of photographic plates is etched slightly by washing with soda to make the emulsion adhere better.

Another auxiliary coat is often added to reduce the formation of halos around the images of bright objects (Fig. 14) a phenomenon well-known from pictorial photography. Scattering of light in the emulsion and reflection at the back of the film or plate generate the halos. The scattered light illuminates the emulsion outside the boundaries of the geometric optical image. These boundaries, therefore, become less sharply defined and the image of a star will be surrounded by an aureole. The direct and the scattered light rays striking the surfaces of the glass or film base at an angle greater than the critical angle will be reflected back into the emulsion (Fig. 15) and will produce a halo around the spot where the primary beam struck the surface of the emulsion. The rest of the rays leave the plate through the back surface of the base. The inner boundary of the halo is quite sharp, the outer diffuse and poorly defined. While the scat-

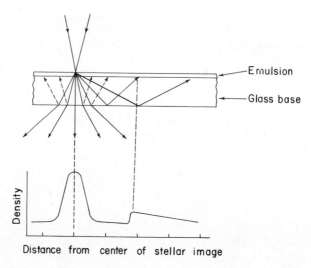

Fig. 15.   Halo formation around a stellar image and densitometer tracing across the image.

tering aureole is only about 10 to 100 microns in diameter, the reflection halo is much larger. Its diameter depends on the thickness of the base and its refractive index. Also, the wavelength of the light and the grain size of the emulsion affect its size.

The diffusion contribution to the halo cannot be prevented. To reduce the halation produced by reflection, the plates are backed by a pigmented coating. This dyed coating absorbs the light to which the emulsion is sensitive. Red-sensitive plates are backed with a dark-green coating that absorbs the red and blue light; yellow-sensitive plates have a red coating that absorbs the blue and green light. The antihalo coating is decolorized during the development of the plate.

The size of the crystalline grains of photographic emulsions varies through a wide range, and it has been found that the different grain sizes and the photographic properties of an emulsion are related to each other.

Formation of the Latent Image

The formation of the latent image in the emulsion by light cast upon it is not fully understood, though a great deal of both experimental and theoretical work has been done on the problem. It seems that the light striking a halide crystal releases electrons at a rate proportional to the intensity of the light. The electrons move at high speed through the crystal until they are caught at certain locations of the crystal structure. These "traps" are presumably impurities containing silver sulfide molecules near the surface of the crystals. Electrons, carrying a negative electrical charge and held in such a place, set up an electrical field of considerable strength and are capable of attracting interstitial positive silver ions, which seem to exist in the crystals.

The electron in the trap combines with the positive silver ion to form a neutral silver atom. The trap then is ready for the capture of another electron and the process begins anew; it might repeat a number of times until a small nucleus of metallic silver is formed. These little silver nuclei in grains all over the photographic plate represent the latent image.

Development of the Latent Image

The silver nuclei in the emulsion act as development centers. The reduction of silver ions to metallic silver by the capture of electrons proceeds much more easily for silver halide grains that have been exposed to light, and hence have these nuclei, than for unexposed grains.

There are two basically different ways of developing the latent image, usually referred to as physical and chemical development. A physical developer contains a reducing agent, often a benzenoid component, a weak acid such as acetic acid, and free silver ions. The weak acid or another suitable compound stabilizes the solution by forming a complex with the silver ions such as silver sulfite. Yet the solution is still able to deposit silver. With age, however, the silver will precipitate out of the solution and be deposited on the walls of the container or on the plate immersed in it. In the process of development the silver is deposited on the nuclei of the emulsion as formed in the silver halide grains by the exposure to light. Thus, the silver forming the visible image on the emulsion is not provided by the silver halide but by the developer. This is true only if a physical developer is used.

A chemical developer does not contain free silver ions; it is a much stronger reducer and does reduce the silver halide grains themselves without adding any silver. The fine structure of the visible image, as revealed by the electron microscope, differs, depending upon whether it is produced by chemical or by physical development.

In practical developing, both kinds of processes may go on. Chemical development may start the reduction of the silver and physical development may complete it. The properties of the developer depend upon its chemical constituents. These constituents are the developing agent, a preservative, an accelerator, and a restrainer.

These agents are quite stable when dry, yet in solution they oxidize quickly. To increase the storage life of the solution a preservative is commonly added. Sodium sulfite, bisulfite, or metabisulfite serve frequently as such. The sulfites oxidize into a sulfate more rapidly than the developing agent, thus retarding the deterioration of the developer. The sulfites are added in different amounts in different commercial developers. A rather common developer is based on metol and hydroquinone. Both components are developing agents; the second improves the keeping qualities of the first. A metol developer when fresh has very much the same properties as a metol-hydroquinone developer, but it deteriorates rapidly, while the oxidization is considerably retarded when hydroquinone is added.

The addition of sulfites can increase the reducing action of the developing agent and act as a solvent of the silver halides. The separation of the grains in the emulsion is increased by the sulfites and a finer grain structure results.

The developing action can be accelerated by adding alkalis, such as sodium carbonate. Developers suitable for high temperatures, often preferred in tropical climates, employ borates and phosphates as accelerators. At high temperatures, the sodium carbonate is carried over from the developing solution and may react with the acid of the fixing bath forming dioxide bubbles ("blisters"). Weak alkalis at high concentration permit better control of the alkalinity than strong alkalis at low concentration.

The developing agent exerts a differential action on the unexposed and the exposed silver halide grains. Development of unexposed grains is undesirable, but happens to some extent and causes what is called background fog all over the plate, veiling the finer details of the image. The composition of the developer is such as to increase the differentiating action between exposed and unexposed silver grains. It has been found by experience that compounds like potassium bromide, potassium chloride, and others increase the differentiation between exposed and unexposed silver halide grains. They are added to the solution as restrainers.

The density of the image—that is, its blackness—depends on a great number of factors, such as the character of the emulsion, the exposure to light, and the degree of development. The degree of development for an exposed emulsion is determined by the composition and dilution of the developer solution and the degree of its exhaustion, the time the emulsion has been subjected to the developer solution, the temperature of the solution, and the degree of agitation of the bath. The degree of development affects the contrast of the developed image. The contrast may become too high if development is carried too far.

Fixing and Washing

The developed emulsion contains the reduced metal forming the image and the residual silver halide, such as silver bromide or silver iodide. They are held in the gelatin, which is swollen with a solution of sodium carbonate or another alkali. In addition, the gelatin layer contains partially oxidized products of the developer. The halides are scarcely soluble in water. To remove them requires a special treatment referred to as "fixing."

Several requirements have to be met by the fixing agent. The stability of the agent must be sufficiently high to achieve complete dissolution of the various silver salts. Also, it must produce complexes stable upon dilution; otherwise, they would decompose during washing. Furthermore, the agent must not affect gelatin or the developed silver grains. Many chemicals are able to dissolve the silver halides, but only sodium or potassium thiosulfates and cyanides fulfill all requirements at a sufficiently low cost. The cyanides are extremely poisonous; therefore, in practice, the thiosulfates, especially sodium thiosulfate ("hypo"), are preferred.

Adequate fixing is often neglected. Even after all visible traces of silver halide grains have disappeared, fixing must not be stopped. At the point of clearing, silver halides are still in the emulsion in an amount sufficient to cause trouble later. The ill effects of insufficient fixing may appear after considerable storage time.

The final washing of the photographic material in water removes the fixing salt, contaminated with dissolved silver components, mostly complexes with thiosulfate. Washing is an important step in the photographic process, and must be thoroughly done. Insufficient removal of silver components will cause stain, mainly in the unexposed or little-exposed areas. Failure to wash out the thiosulfate and its oxidation products causes a slow sulfurizing of the emulsion with time.

The place for processing plates and films after exposure is the darkroom (Fig. 16). Careful planning with proper equipment is an essential requirement for obtaining good results.

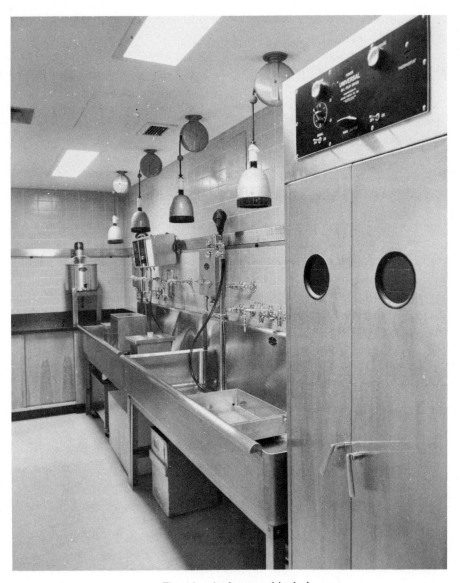

Fig. 16.   A photographic darkroom.

The Characteristic Curve

Photographic material differs in many properties, including relative sensitivity, spectral sensitivity, contrast, granularity, resolving power, and others.

The sensitometric characteristics of an emulsion can be very well represented by plotting the relation between a quality of the photographic image called density and the logarithm of the exposure. The result is known as the characteristic curve of the photographic material.

The strength of the photographic image produced by exposure of the emulsion to light can be expressed in various ways. The mass of silver per unit area would be a measure, but a very impractical one. Another measure is the absorption of light by the developed image. A beam of light is directed through the photographic plate or film, part of it is absorbed and part transmitted. The ratio of the intensity I of the transmitted light to the intensity I_o of the incident light is called the transmission, and the inverse of this ratio the opacity, of the image. Consequently, if the transmission is low the opacity is high. In practice we prefer another term to characterize the image; this is the density, defined as the logarithm of the opacity.

How transmission, opacity, or density can be practically measured will be discussed later (p. 157). These quantities are not exclusively determined by the character of the developed image; they depend also on the type of instrument used for measuring them. If we employ an instrument that illuminates the plate with diffuse light, the values of the transmission, opacity or density will differ from the values obtained if a beam of parallel light is used (Callier effect). The density measured with parallel light is noticeably higher than that measured with diffuse light. The difference between the measurements is related to the size of the grains of the emulsion forming the image; it is smaller for emulsions with fine grain than for those with coarse grain.

The exposure is defined as the product of the intensity of the light and the time during which the plate was exposed to it. Such a characteristic curve is shown in Fig. 17. The curve begins at its lower end with a portion produced by low exposure at a certain minimum density S_o. This is a density found all over the plate, even where it has not been exposed to light; this over-all density is referred to as fog. The amount of fog depends on various factors inherent in the

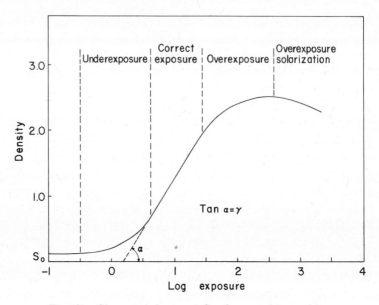

Fig. 17.   Characteristic curve of a photographic emulsion.

development process and the emulsion itself, for example, its age. Older emulsions tend to produce fog more easily than fresh ones. Fog is quite undesirable and can be rather detrimental in astronomical work, particularly if the plates are to be used for photometric purposes. A certain minimum exposure is required to produce a density perceptibly above the density of the fog. The steepness (technically, the slope) of the characteristic curve increases over a certain range of exposure—the range of underexposure of the image, sometimes called the toe of the curve—until the density of the image is built up linearly with the logarithm of exposure. The linear part of the curve is the main portion, showing the range of normal exposure. If possible, exposure times are chosen in such a fashion that the densities produced are obtained in this particular range of the curve. If the exposure is increased further, the density ceases to increase as fast as the logarithm of exposure. Eventually, increase in exposure no longer increases the density. In fact, it might even cause reversal, as additional light action diminishes the density of the image (solarization).

The slope of the straight-line portion of the characteristic curve is a measure of the contrast with which density differences represent intensity differences of the light. The slope of this linear portion is designated as gamma (γ) in the photographic literature. An emulsion with a high value of gamma produces stronger contrast than one with a lower gamma. The tangent of the angle between the straight-line portion of the curve and the exposure axis is the numerical value of gamma. The point where the straight-line section would intersect the exposure axis, if sufficiently extended, is called the inertial point.

As with many other properties of the photographic material, the value of gamma depends somewhat on the color of the light, being, in general, larger for long wavelengths than for short wavelengths. This can be seen easily from a set of characteristic curves for light of different wavelengths (Fig. 18). If nonmonochromatic light is used, the resulting characteristic curve will depend on the distribution of intensity as a function of wavelength.

This has a very practical bearing. On a photograph of a star field, taken for photometric measurements, stars of very different colors (that is, spectral intensity distributions) appear. It is almost impossible to set up a sufficiently large number of characteristic curves to

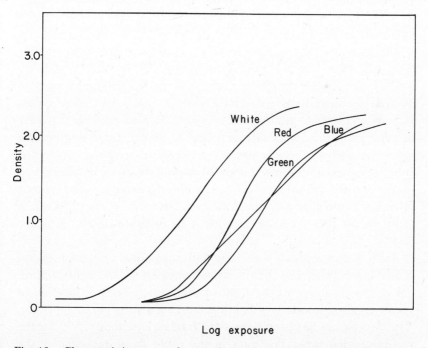

Fig. 18.   Characteristic curves of a panchromatic emulsion for light of different colors.

cover the whole range of stellar colors. By using a common curve for all stars systematic errors might be introduced. Two exposures of different lengths might even lead to different values of the magnitude or brightness difference of stars of different color. Effects of this sort are ordinarily not large, but they need attention if high-precision photometry is aimed at.

Until now, it has been tacitly assumed in our discussion that the density depends exclusively on the exposure, defined as the product of intensity and exposure time, regardless of how the two factors contribute to the product. The situation is actually somewhat more complicated than this, for the photographic reaction to light depends to some extent on how the product is achieved. Thus, for the same exposure, the density produced by a high intensity and a short exposure time is not the same as that produced by low intensity and long exposure time. Correspondingly, one can distinguish between intensity-density curves and time-density curves, depending upon which of the two factors, time or intensity, has been varied. We also have to deal with two somewhat different conceptions of gamma. In photographic photometry, usually the gamma of the intensity-density curve is of interest. These "reciprocity effects" will be discussed in detail later.

The characteristic curve allows us to compare the relative sensitivities of different photographic emulsions. The relative sensitivity depends largely on how the material is used. We might want to compare two emulsions with characteristic curves A and B (Fig. 19). If an object is to be photographed with the least possible exposure to produce a density barely perceptible on the plate or film, the emulsion with the characteristic curve A is more sensitive than B (the plates are used in the region of underexposure). If, however, a higher density is desired, say $D > 1$, and if the exposure can be made adequate, emulsion B is the more sensitive one. Less exposure is required for this material than for A in order to obtain the same density.

Table 1 illustrates the foregoing with numerical values for several Kodak emulsions. The relative sensitivities are given as the reciprocal of the exposure to the light of a tungsten lamp, expressed in meter-candle seconds, which will produce a density of 0.6 above fog. The developing has been done with Kodak Developer D-19 for the recommended time. The data in the table show immediately the proper emulsion with respect to the required order of exposure time,

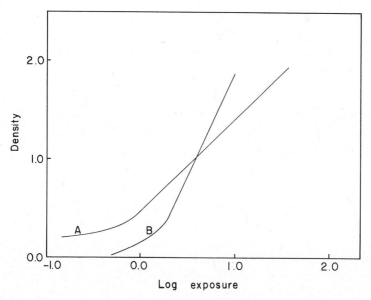

Fig. 19.   Characteristic curves showing the relative speeds of two photographic materials.

if images with moderate density on the straight portion of the characteristic curve are the aim.

As for other light detectors, it is of interest to know the absolute sensitivity of a photographic material, that is, how much energy is required to produce an image of a certain density. The absolute sensitivity of an emulsion can be expressed by a quantity called the

TABLE 1.   SENSITIVITY OF SOME KODAK EMULSIONS FOR THREE ORDERS OF EXPOSURE TIME, RELATIVE TO PROCESS PLATE AND EXPOSURE TIME OF 0.01 SEC.

Plate	Exposure time (sec)		
	0.01	1	10^4
Process	1.0	0.9	0.28
Super Panchro Press	24.0	25.0	2.5
Tri X Panchromatic, B	15.0	19.0	2.5
Spectroscopic 103a-O	18.0	20.0	11.0
Spectroscopic I-O	37.0	35.0	3.2
Spectroscopic IIa-O	4.0	6.6	8.9
Spectroscopic V-O	0.1	0.008	0.006

Source: *Kodak Photographic Plates for Scientific and Technical Use* (7th ed., 1953).

effective quantum efficiency of the emulsion. The number of light quanta that fall upon the emulsion when it is exposed to light is a measure of the energy of the radiation. Each of these quanta could release an electron if it hit a silver halide crystal in the emulsion, provided the energy of the quantum is sufficiently high. As a matter of fact, not all quanta will fall on crystals; indeed, most will not. The ratio of the number of silver grains developed to the number of quanta striking the emulsion during the exposure is the effective quantum efficiency of the material. Actual figures determined for various photographic materials are discouragingly low, somewhere around 0.01 or 0.001, depending on the particular emulsion; that is, only one out of a hundred or a thousand quanta falling upon the emulsion produces a crystal that can be developed.

Reciprocity Failures and Intermittency Effect

We defined the exposure of the photographic material as the product of exposure time and intensity of the light, but it has been stated previously that the two factors are not reciprocal quantities. Different characteristic curves are obtained depending upon whether the exposure time or the intensity is kept constant.

In 1876, Bunsen and Roscoe formulated a law saying that the result of a photochemical reaction depends only upon the total energy employed. The total energy, on the other hand, is the product of intensity and time. From the reciprocal relation between intensity and time the expression is called the reciprocity law. Photographic materials do not obey the law well, and the deviations are known as failures of the reciprocity law or, briefly, reciprocity failures. In astronomical applications exposure times are usually very long and reciprocity failures become quite noticeable and important for selecting the material, as pointed out in the preceding section. It is, therefore, not surprising that a large amount of work on the reciprocity law has been done by an astronomer, K. Schwarzschild (1899).

Information on the reciprocity failures of an emulsion can be obtained by measuring the exposures It required to produce a constant density at different intensities of the light. A graph of the logarithms of exposure It against the logarithm of intensity I, known as the reciprocity-failure curve, is a fair description of the matter. Figure 20 is such a reciprocity-failure diagram. The lines of an

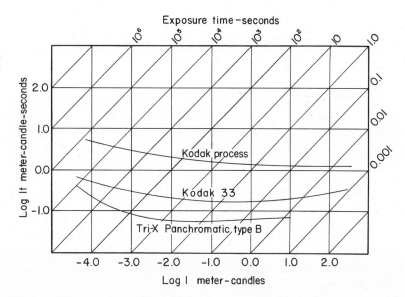

Fig. 20.   Reciprocity-failure curves for some Kodak plates. (Courtesy Eastman Kodak Co.)

angle of 45° are lines of constant time; vertical lines represent constant intensity; horizontal lines represent constant exposure. The intensity increases from left to right, and the exposure times decrease. If no reciprocity failures existed, the plotted curves would be parallel to the horizontal axis of log I, since the same total exposure It, or the same energy would be necessary to generate a certain density, regardless of the intensity level.

For most emulsions the reciprocity-failure curve turns upward at the left. This means that for a very low level of illumination the sensitivity of the photographic material is reduced. All the curves would rise at the right-hand end, if they were extended further. Consequently, each emulsion has an optimum value of intensity at which the greatest photographic effect is obtained for a given exposure.

It has been observed that the reciprocity failures vary with the ambient temperature, as shown in Fig. 21. A series of curves is plotted for different temperatures. With a decrease in temperature, the high-intensity end of the curve rises while the low-intensity end falls at first and then also moves upward. Eventually, the reciprocity failure disappears for a sufficiently low temperature.

As we have seen, the characteristic curve depends on the color

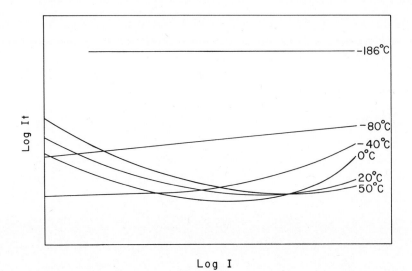

Fig. 21.   Reciprocity-failure curves showing variation with temperature.

or the wavelength of the light and so does the reciprocity-failure curve to some extent (Fig. 22). The curves have the same shape for light of different wavelengths, but they are displaced vertically with respect to one another. If, on the other hand, points of the same density and exposure time are considered, the characteristics of the

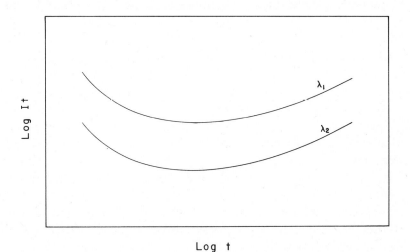

Fig. 22.   Reciprocity-failure curves for monochromatic light of two different wavelengths, λ_1 and λ_2.

reciprocity-failure curve for light of different wavelengths are very nearly the same.

No general rule has been established which relates the shape of the reciprocity-failure curve to other characteristics of an emulsion, such as relative and spectral sensitivity, or grain size. If the photographic material is intended for photometric work, it is highly advisable that the exposure times be equally long in case several exposures are required to set up the calibration curves. Otherwise, the calibration curve obtained for the comparison light sources might not hold for the photographic effects produced by the object under investigation.

Closely associated with reciprocity failures as described is the so-called intermittency effect. A photographic density produced by a continuous exposure is, in general, not equal to the one produced by an equal exposure produced in a number of installments. The intermittency effect is of considerable importance in astronomical applications of photography, because intermittent exposures are quite frequent—in stellar spectroscopy in order to widen stellar spectra; to produce photometric marks by rotating sector diaphragms and in photometric work on stars, when the star images are enlarged by applying a suitable motion to the plate; and on many other occasions.

Figure 23 shows how the intermittency effect depends upon the frequency of the flashes or installments. The curve displays the

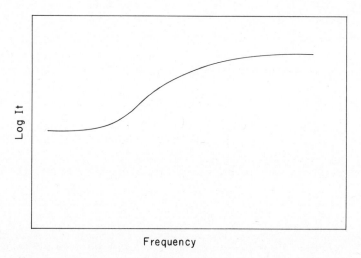

Fig. 23.   Intermittency effect as a function of frequency of flash.

exposure necessary to produce a given density at different flash frequencies. The curve does not necessarily rise with frequency; for another density it might fall. For a particular value of the density, it will run fairly parallel to the frequency axis. From the existence of flat parts of the curve, we can conclude that a continuous and an intermittent exposure of the same energy produce the same density, if the frequency of the installments is below a certain value or above a certain higher frequency. These critical frequencies vary with the intensity of the light. Usually, continuous and intermittent exposures of the same energy produce equal densities if the intermittent exposure is divided into at least 100 installments.

The frequency dependence of the intermittency effect can be easily understood. The flat part of the curve at the low-frequency side means that dividing the exposure into a few flashes does not modify the exposure distribution noticeably from an entirely continuous one. With increasing frequency the curve approaches a value that corresponds to the value required for a continuous exposure of an intensity equal to the average intensity of the intermittent exposure. It has been found that continuous and intermittent exposures are equivalent when the frequency of the installments is approximately equal to the average rate of incidence of light quanta on the sensitive portion of the grains of the emulsion.

A photographic material showing reciprocity failures can produce interesting effects when several exposures of different intensities are imposed upon it. Related phenomena are important in astronomical applications in the form of preëxposure and postexposure.

If a photographic material is very briefly exposed to light of a very high intensity and subsequently to light of low intensity, the two exposures do not add in a simple fashion—an effect discovered by Clayden in 1899. The effect is particularly large if the high-intensity exposure is very short, of the order of 0.001 second or less. The effect sometimes can be studied on photographs of lightning discharges; the images of the lightning sometimes appear reversed as compared with the surrounding background density.

Of greater practical importance is the addition of exposures when a plate is exposed to different fields of stars for comparison purposes. The first field is taken on an unexposed plate, the second on a plate of which the complete area has been exposed to the general background light of the sky. The exposure to the faint background light might be too weak to produce more than just barely noticeable fog or perhaps not even that. The exposure causes nothing more than a "subimage" not yet built up to a developable image. A subsequent exposure will complete the formation of the latent image faster than it would do without the preëxposure. A preëxposed plate seems to have a higher speed.

The action is not the same for exposures to high-intensity light of short duration or low-intensity light of long duration. If one wishes to increase the speed for a photograph to be taken with a long-duration exposure, the plate should be subjected to a very short uniform preëxposure. If the actual photograph requires a short exposure to light of high intensity, a long-time postexposure with a faint general illumination of the plate surface is useful in order to obtain an increase in the speed.

Although the speed of a photographic material can be increased by such methods, it is not recommended that these treatments be applied to plates that are intended for photometric work. The photographic process is not precisely reproducible. Final results depend to an unpleasant extent on factors beyond perfect control. Adding further intermediate steps aggravates the uncertainties and lowers the accuracy of the results. In any case, pre- and postexposures contribute to the plate fog, a highly undesirable condition in photometric work. On the other hand, if extra exposures are used to reduce the length of exposure times for reaching fainter objects only for the purpose of a record, no serious objections to their use exist.

Granularity

If a developed photographic image is examined with adequate magnification, we can see that the silver grains are not uniformly distributed in the emulsion (Fig. 24). Rather they are associated in clumps, frequently overlapping each other. This granularity gives the appearance of graininess sometimes found on photographic prints when they are excessively enlarged.

The granularity of the developed image is an important feature of an emulsion, since it determines to some extent the smallness of image details that are distinguishable. The granularity depends upon many circumstances, including the nature of the developer.

The effects of granularity can be illustrated by scanning a uniformly exposed and developed plate in a microdensitometer. If the size of the scanning beam of light is sufficiently small, the number

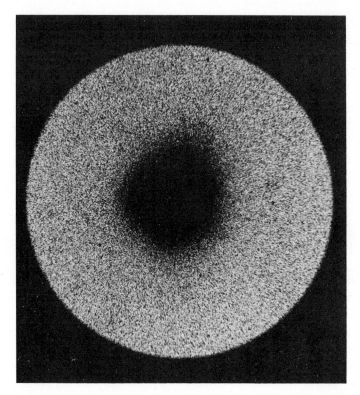

Fig. 24.   Plate grain in the vicinity of a photographic star image, highly magnified.

of grains within the area of the beam on the plate will vary from place to place. Consequently, the transmission or density of the image will vary. If the area of the beam is larger, fluctuations of the transmission will be smaller, since the irregularities of the grain distribution will cancel out. Such microdensitometer tracings from developed plates made with a thin beam are displayed in Fig. 25. Various emulsions obviously have different degrees of granularity. All the emulsions used for the diagram have been treated the same way in development. In general, emulsions with greater granularity are faster. This is not a very strict rule but it is useful to remember when selecting plates. The relation is rather unfortunate for astronomical applications. Because of the faintness of the celestial light sources, plates as fast as possible are desired, but the great granularity that comes with speed restricts the observer to less fast types.

Finer grains can be produced by developing the image to only a low contrast, but this often cannot be tolerated, and other methods have to be used. A general way of decreasing the granularity is the use of developers of low or moderate activity containing silver halide solvents. Such solvents include sodium sulfite and thiosulfate, ammonium chloride, thiocyanates, and organic amines. These agents are sufficiently effective only if added to developers of low activity.

The consequences of plate granularity become quite obvious if one looks at the faintest star images recorded. One may frequently find stellar images that cannot be distinguished from others but that turn out to be spurious when checked against a second plate. These faint images contain only a few silver grains. Such false images are produced by an appropriate arrangement of grains arising from their random grouping. The randomness of the grain distribution also prohibits accurate photometric or positional work on star images at the threshold of visibility.

Turbidity; Resolving Power

Light is scattered when it strikes the silver halide crystals as it enters the emulsion in which the crystals are suspended. When the image projected on the photographic plate is a small area of intense illumination the photographic image generated in the emulsion will consist of a dense center portion surrounded by an area of diminishing density toward the outside. The outer boundaries are the result of the scattering of light at the grains. The size of the outer parts depends on the intensity of the light and the exposure time. If we take a photograph of a star field with a given exposure time, the sizes of the stellar images on the plate represent the brightness of the stars. Indeed, we can determine the stellar brightnesses by measuring the image diameters (p. 163).

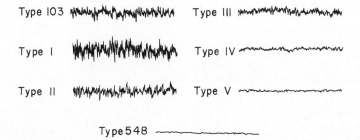

Fig. 25.   Microdensitometer tracings showing the granularity of some emulsions. (Kodak Research Laboratories.)

The tendency of the emulsion to produce such an increase of the image size of bright sources depends not only on the light-scattering power of the emulsion but also on its light absorption. The effects of the elementary processes that lead to spreading of an image with an increase of exposure are called collectively turbidity of the emulsion. The way these processes act and depend on other properties of the emulsion is rather complex.

Turbidity and the inherent contrast of an emulsion determine another important property of photographic materials, their ability to record fine image details in such a way that they can be separated. This ability is known as resolving power. The resolving power of a given material depends on the density; there is a density at which the resolving power is highest. In practice, the resolving power can be determined with the help of line gratings. A series of gratings with different numbers of equal-width black and white lines per millimeter is photographed, and the highest number of lines per millimeter that is resolvable by the eye and with proper magnification is taken as a measure of the resolving power. Table 2 lists some Kodak emulsions and their resolving power determined for an image contrast of 20:1 and a density at which the resolving power is greatest. The table indicates very well the wide range of resolving power various emulsions are able to give.

White light was used for taking the photographs on which the data in Table 2 are based. There is some dependence of resolving

TABLE 2.   RESOLVING POWER OF SOME KODAK EMULSIONS FOR WHITE LIGHT.

Emulsion		Resolving power (lines/mm)
Type 103E		60
Type I-O		60
Tri-X Pan		70
Type II-C		75
Super Ortho-Press		80
Super Panchro-Press		80
Type III-C		95
Process		100
Type IV-C		120
Type V-C		160
Type 548	approx.	500
Type 649	approx.	1000

power on the wavelength of the light, but no simple law or relation between the two and other properties of an emulsion is known, although it seems that the resolving power is in general higher at the short-wavelength end of the visible spectrum and in the ultraviolet than in the yellow and red region.

Many other features bear on the resolving power of an emulsion. The composition of the developer seems to matter little but developing time is important. Usually the resolving power rises rapidly to a maximum, drops slightly, and remains almost constant with further progress of development. A similar behavior is noted with a constant development time but an increase of developer concentration. Increased developer temperatures sometimes reduce the resolution of image details.

Adjacency Effects

The properties of photographic emulsions discussed above are genuine features of the material, though several of them are influenced by the processing. In addition to the inherent characteristics of the emulsion there is a large group of effects that arise during development. They are frequently caused by imperfections of processing, mainly development. Though some of their detrimental actions become more serious by improper handling of the material in development, they cannot always be avoided completely.

The best known of the adjacency effects occurs at the border between a dark and a light area. When a small dense area on the plate, produced by uniform illumination, is surrounded by an area of less density, a microdensitometer tracing may reveal that the density across the darker patch is not quite uniform, and that the boundary is denser than the center (Fig. 26). With a decrease in width of the small dense area the center density may become higher (Fig. 27). Unequal development is the cause for this phenomenon, which is called the Eberhard effect. The developer is used up in proportion to the amount of silver it produces. Part of the development is accomplished by developer contained in the emulsion. This developer diffuses in the gelatin layer. Exhausted developer will move outward from dense areas and less exhausted developer will move into these areas from lighter ones. Close to the boundary line of a dense and a light area strong development will go on at some distance from the dense area, but less strong on both sides of the

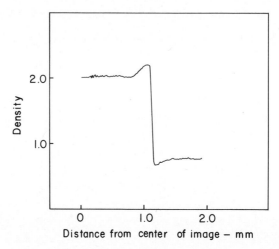

Fig. 26.   Border effect at edge of dense image.

boundary line and still weaker development in the center portion of the dense patch. The smaller the patch is, the less weakened will development of the center be by exhaustion of the developer. Diffusion also occurs on the surface of the emulsion, but its contribution to nonuniform development can be avoided by adequate agitation of the solution. The Eberhard effect can become particularly noticeable if the density changes are large and sudden. Modern

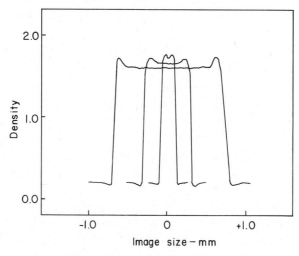

Fig. 27.   Microdensitometer tracings of small areas of different size but equal illumination, showing Eberhard effect.

developers have less tendency to cause Eberhard effects than older and now obsolete types.

Another result of nonuniform development is observed when two small images, as in the case of the images of the components of a double star, lie close together. In the area between the two images, the developer is exhausted faster than in other areas and developing is inhibited where the images touch, whereas it proceeds properly in other areas. The asymmetric development action results in distortion of the shape of the images of the two double-star components and erroneous measurements of their angular distance. The separation turns out to be too large. The effect is known as the Kostinsky effect; it might occur also on a spectrogram with close multiple emission lines.

Geometric distortion of close images of stars also results while the plate dries after processing. The gelatin becomes tanned during development, and the amount of tanning is greater in areas where the silver density is higher. The more heavily tanned sections dry faster, and, since the gelatin tends to contract when it dries, the tanned areas shrink and pull inward neighboring areas that are still wet. Thus, the strong image of a bright star will attract toward itself the images of faint stars in the vicinity. Components of a close double star might appear at a smaller angular distance than they really are. Photographic material should be dried carefully whenever accurate measurements of positions are intended. Addition of a wetting agent to the rinse water is advisable, because it reduces the surplus water on the film or plate surface.

The adjacency effects, including some minor ones not mentioned here, cannot be well controlled, although vigorous agitation of the solution reduces some of them; effects on the density will exceed 5 percent in particular cases only. Shifting of images should not exceed 10 microns, if proper care in drying is taken.

Spectral Sensitivity and Hypersensitization

The sensitivity of the photographic emulsion varies with the color of the light to which it is exposed. Pure silver bromide does not respond to light of longer wavelengths than about 4600 A, and pure silver chloride becomes insensitive above 3900 A. Addition of silver iodide shifts the limit of sensitivity to almost 5000 A. In 1873, Vogel discovered that the addition of a dye to a silver bromide emulsion confers sensitivity to that section of the spectrum that is absorbed by the dye. Not all dyes perform this change in sensitivity equally well. A great number of sensitizing dyes have been found by now, most of them belonging to groups known as cyanines, merocyanines, and carbocyanines. The advent of these dyes has made it possible to produce photographic reactions in emulsions with light of a wavelength that is otherwise too long to trigger the elementary processes of image formation.

By selecting and combining the dyes to be added to the emulsion, it is now possible to cover the electromagnetic spectrum from the γ-ray and x-ray region at the short-wavelength end through the visible to infrared radiation of almost 13,000 A. An extension of the sensitivity range to even longer wavelengths is not likely to be feasible, because dyes that absorb radiation of such long wavelengths would be subject to thermal decomposition; also water absorption of long-wavelength radiation may limit further extension.

A summary of the various spectral sensitizings obtainable for Kodak emulsions is given in Fig. 28. The emulsions can be sensitized for the various spectral regions as indicated in the diagram. The emulsions themselves differ by properties such as relative sensitivity, graininess, contrast, and others. In Fig. 29 are shown samples of sensitivity curves of emulsions having various combinations of type and sensitizing class. The sensitivity whose logarithm is plotted

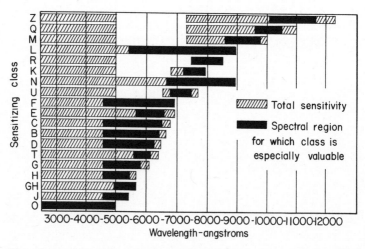

Fig. 28.   Spectral sensitivity region of Kodak spectroscopic plates. (Courtesy Eastman Kodak Co.)

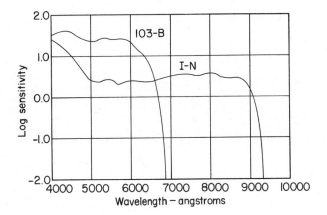

Fig. 29.   Spectral sensitivity curves of some Kodak emulsions. (Courtesy Eastman Kodak Co.)

is defined as the reciprocal of the number of ergs per square centimeter required to produce a density of 0.6 above fog, the plate having been developed as recommended by the manufacturer.

The spectral sensitivity curves of various photographic materials can be modified further by employing color filters. By a suitable combination of filters and emulsions, records produced by light of any desired limited wavelength range can be obtained.

Emulsions with a particularly high sensitivity are less stable than those with a lower sensitivity. They have a stronger tendency to develop background fog when stored. This tendency prohibits manufacture of emulsions exceeding a certain sensitivity, because they could not be kept for a sufficient length of time. Emulsions are usually adjusted in such a fashion that the rate of fog growth in storage is very low, a procedure that can be followed only at the expense of sensitivity. However, it is feasible for the user to adjust the chemical condition of the emulsion of the photographic material by various treatments in such a way that the sensitivity is increased. To prevent fog development this process, called "hypersensitization," is carried out only shortly before the exposure to light. Hypersensitization of an emulsion increases the sensitivity in the spectral region for which the material has already been optically sensitized. Hypersensitization is essential for infrared sensitive emulsions.

A great variety of treatments capable of achieving this increased sensitivity have been found. Bathing the plates for a few minutes in cold water serves the purpose for many types of emulsion. A more efficient procedure for hypersensitization, particularly for red and infrared material, is bathing in a dilute ammonia solution. Rapid drying afterward insures the least increase in background fog. Plates or films should be hypersensitized only if speed requirements make it mandatory. The additional treatment enhances the danger of mechanical damage and contamination of the emulsion. In photometric work, hypersensitization, if not uniform over the area of the plate, may lead to measuring errors.

3

Telescope Optics

In spite of its age of some three and a half centuries, the visual telescope (even in a form closely resembling the invention of Galileo) is still an effective astronomical instrument. Emphasis in recent years has been on telescopes specifically designed for photography, but visual instruments are still, and will probably continue to be, the best telescopes for certain problems. For example, for measuring close double stars, observing detail on the surface of planets, and recording sunspots the visual telescope remains supreme. In addition, a visual telescope is generally a necessary fixture on large photographic telescopes for finding objects. Accordingly, visual telescopes are treated here in some detail. Before discussing the telescopes themselves, we must study the instrument that they serve—the human eye.

The Human Eye

When a telescope is used for visual observing, the eye becomes part of the optical train. The optical parts of the eye are relatively simple and together are similar to a photographic camera. Figure 30 is a diagram of the eye. Roughly, the eye consists of a lens, a light-tight enclosure which also contains the medium behind the lens, and a light-sensitive film, the retina.

The lens of the eye is not just a simple lens, however, but consists of several elements—the cornea, the aqueous humor, the crystalline lens, and the vitreous humor. The most important of these actually is the cornea, for this element performs most of the focusing of the entering light. The crystalline lens is the variable element in the system which permits the eye to focus on both near and far objects. It is a fibrous jelly whose shape can be changed by a sphincter muscle around its periphery. As a human eye ages, the crystalline lens slowly loses this power of change of shape; the eye loses its accommodative faculty and its focal length becomes fixed.

The aqueous humor is a saline solution, while the vitreous humor is a jelly. These fluid media, on either side of the crystalline lens, permit it to change its shape.

The retina at the rear of the lens contains the light-sensitive elements of the eye. These elements are in the shape of minute rods and cones extending from the retinal surface. The resolving power of the eye is largely determined by the granular structure formed by the rods and cones of the retina. The distribution of the rods and cones is not uniform over the retina. The cones are concentrated in a small area near the intersection of the optic axis of the lens with the retina. This region, called the *macula lutea*, is a slightly depressed spot of the retinal surface a little to one side of the optic axis. In the *fovea centralis*, at the center of the macula lutea, only cones are present and vision is most acute at this point. In this region two point images as close as 10 microns can be resolved. This is about the separation of a cone from its neighbor. In terms of angle the eye can

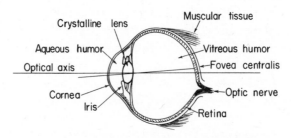

Fig. 30.   The human eye as an optical instrument.

separate points as close as 1 minute of arc. The diameter of the fovea centralis is only 0.25 mm and consequently the field of most distinct vision is only about twenty-five times greater than the minimum angle that the eye can resolve. Indeed, this region of distinct vision is smaller than the moon appears to the unaided eye, so that one must shift his vision to see different parts of the moon distinctly.

Recent research shows that the eye is never still but is constantly oscillating through an angle slightly greater than that between cones at a rate near 60 cycles per second. If this oscillation is stopped or the image that the eye is observing is otherwise made stationary on the retina, the eye fails to see the image after a short time. The oscillation may also serve to eliminate the grain structure in the image that would be caused by the mosaic nature of the retina.

An important part of the eye is the iris diaphragm, immediately in front of the lens. Its function is precisely the same as that of the iris of a camera lens—to regulate the amount of light entering the lens. The human iris is controlled by a sphincter muscle which can vary the diameter of the iris from about 1 mm for bright illumination to 8 mm for faint illumination. These diameters are important to keep in mind, for they enter into the design of a visual telescope. The diameter of the iris for comfortable illumination is about 2 mm. Maximum acuity also occurs at about this diameter. At smaller diameters acuity is limited by diffraction and at larger diameters by aberrations of the lens of the eye. When the iris has a diameter of 2 mm the three factors, aberration, diffraction, and the retinal structure, all help to limit the acuity.

Though the rods do not give as sharp an image as the cones, they are more sensitive to faint light. Since the rods are outside the region of maximum acuity, one must use averted vision (looking beside an object) when seeking the faintest objects visible. The rods depend on a pigment called visual purple, a derivative of vitamin A, for their functioning. A lack of vitamin A in the body definitely reduces night vision. After the retina is exposed to bright light, the supply of visual purple in the eye is used up and the eye cannot see faint light again until the visual purple has been replenished. The process of renewing the supply of visual purple, called dark adaptation, takes 30 minutes or so. The variation in the amount of visual purple present allows the eye to function throughout a light range of 10 billion times. Rapid readjustment to small changes in light level is made by the iris but its range is only about sixteenfold.

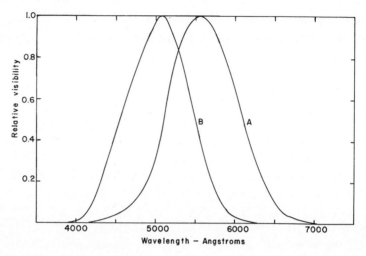

Fig. 31.   The relative sensitivity at various wavelengths of (*A*) the light-adapted eye and (*B*) the dark-adapted eye.

The spectral sensitivity of the eye depends on the light level. The dark-adapted eye has its maximum sensitivity at 5100 A, as shown in curve *B* of Fig. 31. The eye under normal illumination has its maximum sensitivity at 5500 A, as curve *A* shows. The shift of the sensitivity toward the blue as the illumination is decreased is referred to as the Purkinje effect.

The Refracting Telescope

A visual telescope consists of two essential parts—the objective lens, which forms an image of the observed object just as the lens of the eye does, and an eyepiece, which permits the eye to view this image. The arrangement of an objective and eyepiece is shown in Fig. 32, where simple lenses are shown in place of the compound lenses that usually constitute the two elements.

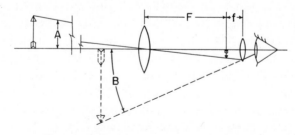

Fig. 32.   Magnification in a visual telescope.

The visual telescope has two functions—to increase the amount of light collected and concentrated on the retina of the eye so that fainter stars may be seen, and to magnify the image falling on the retina so that more detail in the image may be seen. We shall discuss the latter function first.

The distance between a simple lens and the image that it forms of a distant object is called the focal length of the lens. Since the size of the image formed is proportional to this focal length, objective lenses with long focal lengths are frequently desired for their large images. When a compound lens is used, a quantity called the effective focal length, or often just the focal length, is defined. It is the focal length of a simple lens that would form an image of the same size as the compound lens does.

Figure 32 shows how magnification is produced by a telescope. The arrow at the left of the drawing is the object, and if it is far away it is imaged by the objective at the focal point. Its image is magnified by the eyepiece so that it appears to the eye as if the large dotted arrow were actually seen. The object subtends the angle A at the telescope objective, and also at the eye since the distance from the objective to the eye is negligible compared with the distance to the object. The eyepiece magnifies A so that it appears to the eye as the angle B. The magnifying power of the telescope is defined as the ratio B/A, and the position of the eyepiece is chosen so that this ratio is equal to the ratio of the focal lengths F/f. The magnifying power of a telescope can be changed simply by substituting an eyepiece with a different focal length. The choice of the magnifying power will be treated later.

Objective Lenses

The most important lens of the telescope is the objective. On it depend the highest magnifying power that may be used to advantage and the magnitude of the faintest stars that may be seen. Both of these properties are influenced largely by the diameter of the objective. As a result telescopes are usually characterized by the diameter of the telescope objective. The largest refracting telescope is the 40-inch at the Yerkes Observatory.

The function of the objective is to form as sharply defined an image of a star as is possible. As was pointed out in Chapter 1, the wave nature of light through the diffraction effects of the lens aperture establishes the limit for a given lens beyond which it is impos-

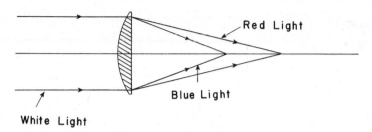

Fig. 33.   Chromatic aberration of a simple lens.

sible to produce a more sharply defined image. It is expected of visual telescopes that the limitation of diffraction will be the only limit to resolution. In an improperly made or designed lens, however, there may be a host of other things that increase the size of a star image above the size resulting from diffraction.

Even a well-made simple lens has several limitations to the resolution more severe than the limit imposed by diffraction. Sharpness-limiting properties other than diffraction are called *aberrations.*

The most important of the aberrations affecting a simple or singlet lens is *chromatic aberration,* illustrated in Fig. 33. This aberration is due to the variation of the refractive index with wavelength, as described in Chapter 1. The result is that an image produced by blue light is formed closer to the lens than one produced by red light; hence white light gives a more or less blurred image. The best focus is obtained at a plane perpendicular to the axis of the lens between the red and the blue images. At this plane, the red and blue images of a point are coincident disks, which together are called the *circle of confusion.* The diameter of the circle of confusion is independent of the focal length of the lens, while the size of the image is proportional to the focal length, as was previously mentioned. As a result, the best definition is obtained by using the longest focal length that is practical. Before the invention of the achromatic objective some telescopes with monstrous focal lengths were constructed; the one shown in Fig. 34 is an example.

Another aberration of the simple lens is *spherical aberration.* A lens with spherical surfaces is much easier to construct than ones with other shapes, but such a lens does not bring all rays to a common focus. Rays from the edge of the lens are focused nearer to the lens than those passing near the middle, as Fig. 35 shows. Spherical aberration is also smaller in a lens of longer focal length. It can be

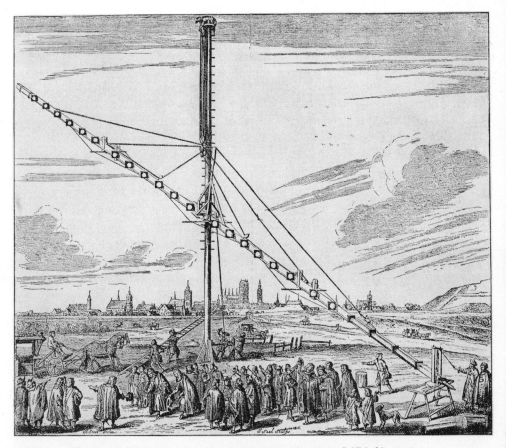

Fig. 34. The 150-foot focal-length telescope of Hevelius. (Courtesy *Griffith Observer*.)

eliminated by grinding and polishing nonspherical surfaces, but another method is customarily used.

Both chromatic and spherical aberrations may be greatly reduced by making a doublet lens, that is, a combination of two lenses having glasses of different composition. If a converging, or positive, lens of crown glass is combined with a diverging, or negative, lens of flint

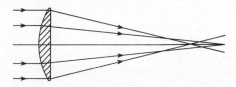

Fig. 35. Spherical aberration of a simple lens.

glass, the dispersions can be made to cancel without complete cancellation of refractive power. Sir Isaac Newton believed that the dispersive power of a glass was proportional to its refractive power, namely, $(n - 1)$, and from this false idea he was led to the conclusion that it was impossible to construct an achromatic lens. C. M. Hall discovered about 1733 that this was not the case, and John Dollond manufactured achromatic lenses about 1759. Figure 36 shows such a combination lens, or doublet.

The chromatic aberration is not canceled for all wavelengths, however, but only for a narrow range. The range in which the focal length is nearly constant is chosen to coincide with the region of maximum sensitivity of the eye, that is, in the green at 5500 A. The variation of the focal length with wavelength of the Yerkes 40-inch refractor is illustrated in Fig. 37. This example is an extreme case but the residual color error or secondary spectrum for doublet lenses larger than 3 inches is not negligible. The achromatic lens reduces the total range of the focal length through the visible to 5 percent of that for a singlet lens. Special glasses may reduce it even further, but such glasses have not been found to be as stable or cannot be produced in as large disks as the more usual glasses. As Fig. 37 shows, the violet and the red are brought to focus at points farther from the objective than the green. If the telescope is focused to give a sharp image of a star in the green, the red and violet images will have a disk shape. The result is a purple halo about the star image. As a consequence of the reduction of the chromatic aberration to 5 percent of that of the uncorrected lens, the telescope will have no more unwanted color in its image than one with a singlet lens of 20 times the focal length. The monstrous focal lengths used in the past are no longer necessary.

Another parameter used to classify objectives is the f-number, which is the ratio of the focal length to the diameter of the objective. For example, the 40-inch telescope has an f-number of $f/18$, since its focal length is 60 feet. Visual refractors for astronomical use generally have f-numbers between $f/10$ and $f/20$. Where planetary

Fig. 36.   The achromatic objective as made by Fraunhofer.

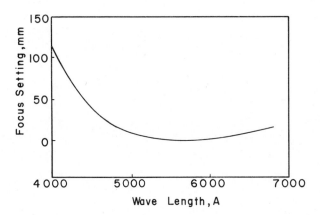

Fig. 37.   The residual chromatic aberration of the 40-inch refractor.

observing is intended, as large a value as can be managed should be chosen, even as high as $f/30$.

By properly choosing the radii of the front and back surfaces of each element of the doublet, the spherical aberration may be reduced. If the radii of the two elements are so chosen that the focal length for the outer rim of the combination is the same as the focal length of its central zone, then the zonal ring at 0.7 of the radius of the objective will have a focal length that is shorter than that of any other zone. Such residual spherical or zonal aberration is illustrated in Fig. 38. Curve (a) shows the spherical aberration of a simple double-convex lens, and curve (b) the zonal aberration for a corrected doublet. Zonal aberration may be reduced still further by polishing glass from the intermediate zones of one of the surfaces so that all zones of the doublet have the same focal length; in other words, a surface of one of the lenses is made slightly nonspherical. A lens from which the makers, Alvan Clark and Sons, took great pains to eliminate the residual spherical aberration is the 24-inch objective of the Lowell Observatory. The variation of the focal length of this lens along a radius is shown in curve (c).

Spherical aberration may be minimized by many different pairs of lens elements. Elements having the same radius of curvature of adjacent surfaces may be cemented together with Canada balsam if the diameter is not over 7.5 cm. If objectives of larger diameters than this are cemented, strains occur with changing temperature and accordingly large cemented objectives are not recommended. Lenses have sometimes been designed with the flint-glass element

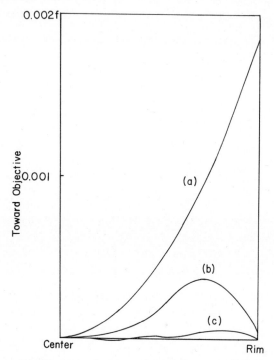

Fig. 38. The residual spherical aberration of lenses: (*a*) the spherical aberration of a double-convex lens; (*b*) the zonal aberration remaining in a doublet with spherical lenses; (*c*) the aberration of the 24-inch Lowell Observatory refractor after local polishing to remove the zonal aberration.

toward the front. These have the disadvantage that the more unstable and softer glass is exposed to weathering and abuse.

Eyepieces

The other optical part of the visual telescope, the eyepiece, is nearly as important as the objective for obtaining sharp definition and suitable magnification. Eyepieces are not generally a fixed part of the telescope but are interchangeable. In this way a range of magnifying powers can be obtained with one objective.

The eyepiece functions as a magnifying lens. Although the image formed by the objective can be observed by the eye without any further lens, this is never done because of the low magnification and small field. Usually the eye cannot focus comfortably on an object closer than about 25 cm. An eyepiece permits the eye to see the image at a convenient distance when the object is very close to the eye and hence subtends a large angle. The object for the eyepiece is the image formed by the objective. The eyepiece is focused to produce parallel light from the close-up image of a star. The eye is then focused at infinity and consequently is relaxed. Slightly higher magnification may be obtained by focusing the eyepiece so that the magnified image appears at the nearest distance on which the eye can focus comfortably as in Fig. 32, but over a long period this strains the eye and should not be done.

Besides producing parallel light from the image formed by the objective, the eyepiece images the objective. The image of the objective, shown at the right of Fig. 39, is called the *exit pupil* or *Ramsden circle*. The exit pupil may actually be seen if the telescope is pointed to the daytime sky and the eye placed a foot or so behind the eyepiece. A bright disk of light can then be seen. If the head is moved slightly from side to side the disk will appear to be fixed about 2 cm from the eyepiece. To find its exact location, place the point of a pencil between the eye and the lens and move the point along the line of sight while moving the head from side to side. A position will be found for the pencil such that the disk and the pencil do not move relative to each other. The pencil point then lies exactly at the location of the disk. The distance of the Ramsden circle from the eyepiece is called the *eye relief*.

All light that goes through any part of the objective also goes through the corresponding part of the exit pupil. Hence if all the light collected by the objective is to enter the eye, the iris of the eye must be at least as large as the exit pupil, and the best location for the iris is at the exit pupil. This location is made possible by the eye relief.

The requirement that the exit pupil be at least as small as the iris of the eye establishes a lower limit to the magnification that can be used while still making use of the full aperture of the telescope

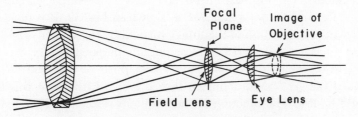

Fig. 39.   Illustrating the effect of a field lens.

lens. The diameter d of the image of the objective is equal to the product of the diameter D of the objective and the ratio of the focal length of the eyepiece to that of the objective, that is,

$$d = D \times f/F.$$

From this equation the ratio of diameters D/d is equal to the magnification F/f. The diameter of the exit pupil may be measured by holding a scale up to it while the telescope is pointed skyward. This is a way to determine the magnification of a telescope when an eyepiece of unknown focal length is used.

The formula for the diameter of the exit pupil shows that, if the focal length of the eyepiece is too large, the diameter of the exit pupil will be larger than the iris of the eye. When this is the case, not all of the light gathered by the telescope objective is utilized. The brightest images are produced by the telescope when the diameter of the exit pupil is just equal to that of the iris. This advantage should be sought when looking for the faintest objects that can be seen with an instrument. The magnification satisfying this condition is obtained from the second formula by putting the normal maximum iris diameter, 8 mm, for d. As a rule, the magnification yielding the brightest images is three times the number of inches in the objective diameter.

Magnifications greatly in excess of those given by the foregoing rule are used when plenty of light is available. In observing planets and double stars, enlargements as high as 50 times the objective diameter are sometimes used. On any given night, however, a magnification is found where the maximum amount of detail may be seen. This depends on the steadiness of the atmosphere. The air is in constant motion or turbulence, causing telescopic images to dance and shimmer. If the dancing is rapid, they may appear blurred but stationary.

Eyepieces of the simple-lens type are rarely used in astronomical instruments. The simplest improvement over the simple magnifying lens is the addition of a *field lens,* as shown in Fig. 39. It is placed at or near to the position of the image formed by the objective. At this point it has negligible effect on the magnifying power. It serves, however, to divert light from the outermost parts of the field of view toward the *eye lens* and thence into the eye. Otherwise this light would be lost. The function of the field lens is to extend the field of view of the eyepiece; hence its name.

The outer part of the field of view permitted by an eyepiece may not be sharply defined. An aperture is usually placed at the focus of the eyepiece to limit the field to the sharply defined part. The sharply defined field is expressed by the angular cone visible to the eye. Eyepieces have fields of view ranging from 20° to 70°.

The field of view of the entire telescope is found by dividing the field of the eyepiece by the magnifying power. Thus with a 70° eyepiece field and a magnifying power of 100 the field is 0.7° and would include a little more than the entire moon.

The field lens is seldom placed exactly at the focus of the objective. It is usually displaced somewhat in front of or behind the focus so that any dust particles on the lens do not appear in sharp focus as seen with the eyepiece. In addition, the actual focal plane is free for the inclusion of cross hairs if these are desired.

The field lens in many eyepieces is used to correct chromatic aberration of the eyepiece as well as to serve its primary function. The chromatic aberration that is important in an eyepiece is of a different variety from that which concerned us in the objective. There we were worried about the location of the images of different colors, that is, whether they all fell in the same plane. This form is called longitudinal chromatic aberration. In the eyepiece the important requirement is that the magnification be the same for all colors. When this is not the case the lens has lateral chromatic aberration. This defect can be eliminated by a combination of two separated lenses of the same kind of glass.

The Huygens eyepiece shown in Fig. 40(a) is free of lateral aberration. Two plano-convex lenses are used; their focal lengths and separation are given in the figure for an eyepiece whose focal length is 1 inch. For any other focal length these dimensions should be changed proportionally. The Huygens eyepiece may be easily constructed from suitable lenses obtained from optical supply houses.

Since the focal plane of the Huygens eyepiece falls between the lenses, the eyepiece cannot be used as a magnifier and for this reason is called a negative eyepiece. Cross hairs may be mounted at the focal plane but this is rarely done since only the eye lens is used to magnify them and their definition is not good. The Huygens eyepiece performs well with objectives with f-numbers as small as f/8 and the well-defined field of view is 40°.

The Ramsden eyepiece [Fig. 40(b)] also is formed of two plano-convex lenses, but here they have the same focal length. Instead of

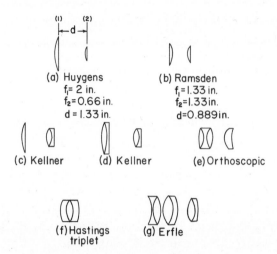

Fig. 40.   Some of the eyepieces that are used in astronomical telescopes.

being separated by exactly their focal length (which would eliminate the lateral color completely) they are spaced a little closer than
this so that the focal plane is outside of the combination and does
not coincide with the field lens. The Ramsden eyepiece may be used
as a magnifier and is quite suitable for use with cross hairs in precision measuring instruments. Since it is not completely corrected
for lateral color, it images well a field of only 35°. On the other hand,
it can be used with objectives with small f-numbers ($f/7$).

The Kellner eyepiece [Fig. 40(c)] combines the advantages of the
Ramsden and achromatization. This is accomplished by making the
eye lens a cemented doublet of flint and crown glasses. Sometimes
the field lens is achromatized also, as in Fig. 40(d). The Kellner is
a good eyepiece to use when a cross hair is required. The sharply
defined field of the Kellner is 45° and it can be used with an $f/6$
objective.

A disadvantage of these three eyepieces is the short eye relief,
which is a result of the field lens. A longer eye relief is obtained by
eliminating the field lens, as is done in the Abbe orthoscopic eyepiece [Fig. 40(e)]; a large field may be obtained by making the lenses
large. With the four elements this lens has a highly corrected color-
less field of 30° and can be used with an $f/6.5$ objective. Another
eyepiece without a field lens and hence with long relief is the Hast-
ings triplet [Fig. 40(f)]. It has a field of 30° and may be used with
an $f/6$ objective.

One of the finest eyepieces is the Erfle [Fig. 40(*g*)]. It has a very sharply defined field of 65° and may be used with an *f*/6 objective. It makes a very excellent eyepiece for viewing the entire moon at moderate magnification, for comet-seeking, or for examining nebulae where the lowest magnifying power consistent with the requirement set by the iris diameter is sought.

The inclusion of cross hairs in eyepieces has already been mentioned. For some auxiliary instruments the cross hairs may be moved by a precisely calibrated screw. One such instrument, which is used for measuring close double stars, is the bifilar micrometer (Figs. 41 and 42). As the name suggests, this device has two parallel threads. One thread is moved parallel to itself by the precision screw attached to the calibrated drum. The other thread is slightly above the movable one so that the two may pass but not enough to make it out of focus in the eyepiece. Usually there is a third thread which is fixed at right angles to these two. The box that supports the fixed and movable threads is itself movable by a screw so that both threads may be moved simultaneously. In addition to this motion, the box may be rotated about the optical axis of the telescope and the angle read accurately on a divided scale.

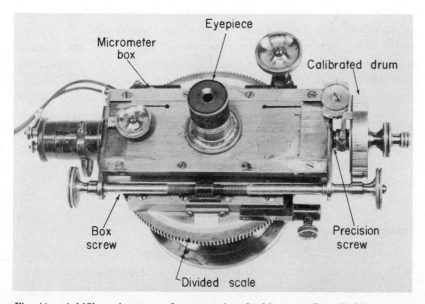

Fig. 41.   A bifilar micrometer for measuring double stars. (Lowell Observatory photograph.)

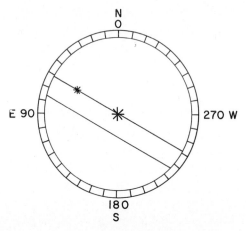

Fig. 42.   Illustrating the use of the bifilar micrometer. Here the fixed cross hair is adjusted for determining the position angle of the double star.

The procedure for measuring a double star is first to rotate the micrometer box so that the fixed thread passes through the two star images, and then to read the circular scale. The average of several such settings will give an accurate measure of the position angle of the star (see Fig. 42). The head is then set at 90° from this angle, and the stellar separation is measured by setting the fixed thread on one star with the box screw and adjusting the movable thread to the other star with the precision screw. After the drum position is read, the fixed thread is moved to the second star by the box screw and the movable thread is moved to the first star, meanwhile passing the fixed thread. It will be seen that the movable thread has been moved through twice the separation of the two stars. From the scale of the telescope (number of seconds of arc corresponding to a 1-mm displacement in the focal plane) and the distance through which the thread has been moved the angular separation of the star is obtained.

Reflecting Telescopes

Astronomical telescopes frequently employ mirrors instead of lenses. For a given aperture it is generally much cheaper to construct a mirror than a lens. A mirror of concave *parabolic* shape (Fig. 43) is used. (Astronomers frequently use the term "parabolic" where the word "paraboloidal" is actually meant. A parabolic mirror has the

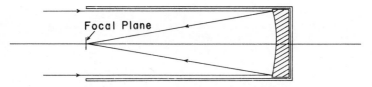

Fig. 43.   The parabolic mirror and the prime focus.

shape of a parabola rotated about its axis. Similarly, a hyperbolic mirror has the shape of a hyperbola rotated about the axis through the two foci and an elliptical mirror is a portion of the surface formed by rotating an ellipse about its axis.)

A property of the parabolic mirror is that incoming rays of light parallel to its axis are reflected to a common focus. In constructing a telescope mirror the optician uses special skills to make the actual mirror figure parabolic in shape rather than spherical, which is much easier to make. By this stratagem one of the problems of the objective lens, spherical aberration, is eliminated. Another great advantage of a reflecting mirror is the absence of chromatic aberration. All wavelengths of light are brought to the same focus.

The prime focus, located on the axis of the telescope, is rarely used for visual work but is used occasionally for photography, spectroscopy, and photometry. Indeed, in the 200-inch reflector, the observer rides the telescope in a cage that is supported right on the axis of the telescope at the prime focus (Fig. 44). In Herschel's form of the telescope, however, the mirror is inclined to divert the prime focus to the side of the tube where it may be directly viewed with an eyepiece. This may be done when the telescope has a large *f*-number so that aberrations which result from this inclination are not severe. Sometimes in the Herschel telescope the mirror is a portion of a parabolic mirror that is cut off-center after the whole mirror is made. Aberrations in this form are less than with the inclined mirror.

The arrangement of the reflector built by Sir Isaac Newton about 1672 and now called the *Newtonian* type is shown in Fig. 45. It is probably the most popular kind of reflecting telescope for visual use. An auxiliary flat mirror, called the *diagonal* or *Newtonian flat*, is placed on the axis of the telescope at an angle of 45° so as to divert the focus to the side of the telescope tube. Its size is such that it does not obstruct much (about 10 percent) of the incoming light. Another 10 percent of the light is lost by absorption at the mirror surface. The advantages of accessibility of the focus at the side of the tube

Fig. 44.   The prime-focus cage of the 200-inch Hale telescope. (Mount Wilson and Palomar Observatories photograph.)

far outweigh the slight loss of light. The diagonal is supported within the tube by means of three or four equally spaced struts, called a spider. Although this structure blocks a negligible amount of light, it does cause a four or six-spiked pattern of rays about star images (see Fig. 69). The spikes are due to the diffraction of light by the struts.

Another form of the reflecting telescope, the *Cassegrainian* (Fig. 46), uses a convex hyperbolic mirror on the axis of the telescope, which both diverts the converging light through a hole in the primary mirror to a focus behind the mirror and magnifies the image as well. The focus is particularly convenient for observing and also for mounting attachments. Cassegrain instruments usually have *f*-num-

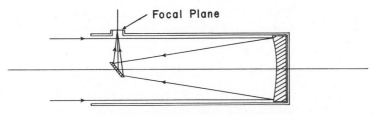

Fig. 45.   The Newtonian telescope.

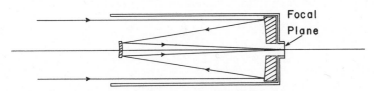

Fig. 46.   The Cassegrainian telescope.

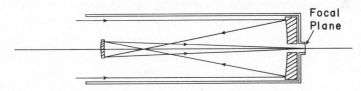

Fig. 47.   The Gregorian telescope.

bers between $f/10$ and $f/30$. Sometimes the converging light instead of passing through a hole in the primary mirror is bent to the side of the tube by a third, flat mirror as in the Newtonian telescope. This arrangement is called the modified or bent Cassegrainian telescope.

The Gregorian telescope (Fig. 47) has a concave elliptical secondary mirror which reimages the prime focus through a hole in the primary mirror. It requires a longer tube than the Cassegrain telescope, and consequently a larger dome is needed. This results in a substantially larger cost for the instrument and so it is almost never made professionally. The secondary mirror is much easier to make than the Cassegrain mirror, and the Gregorian finds some favor with amateurs for this reason.

The *coudé* telescope is mainly a professional type for photographic or spectroscopic use, but is discussed here for completeness of text. The coudé takes several forms, depending on the type of mounting of the instrument. Figure 48 shows a common form. It has a Cassegrain secondary mirror and a flat mirror that rotates through half

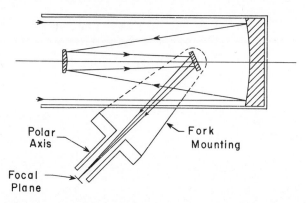

Fig. 48.   One form of coudé telescope. The focus at the end of the polar axis is in a fixed position.

the angle that the delination axis does so that the light will be reflected through the hole in the polar axis, thus providing a focus that does not move about with the telescope. The instrument is, therefore, ideal for use with very heavy or bulky auxiliary equipment, such as high-dispersion spectrographs. As the telescope turns on the polar axis, however, the image rotates. Plate holders may be attached to the polar axis so as to rotate with the image, or a *Dove prism* may be used to rotate the image in the opposite direction to keep it fixed in position on the slit of a stationary spectrograph.

A visual telescope with a fixed position for the focus is the Springfield (Fig. 49), named after the group of amateur telescope makers at Springfield, Vermont where it was invented. It is essentially a Newtonian telescope but has an additional diagonal mirror to divert the light up the polar axis to a focus at a fixed position.

Performance of Visual Telescopes

One of the chief uses of a visual telescope is to look at faint stars. The faintest star that can be seen with a telescope, as long as the exit pupil matches the pupil of the eye and the telescope loses little light by absorption, depends only on the area of the objective, which is proportional to the square of its diameter. The diameter of the iris (the eye is an objective too) is ⅓ inch (8 mm) at its largest. With a telescope of diameter D inches one can see stars that are as much as $9D^2$ times fainter than those just visible to the naked eye. The unaided eye can see stars only to magnitude 6.2. A 1-inch telescope will make visible stars of magnitude 8.8. Each increase of ten in telescope diameter will make visible stars 5 magnitudes fainter. Thus with the 100-inch one can see stars of magnitude 18.8. All this is on the assumption that the sky is perfectly clear, the atmosphere is not turbulent (good seeing), the telescope objective has high transparency, and the sky is not illuminated by city lights.

The foregoing analysis for stars does not apply to extended objects. Nebulae that can already be recognized as diffuse objects will not appear brighter to the eye no matter how much the size of the telescope is increased. This results because, although a larger objective collects more light, the light is spread over a larger part of the retina, as the magnification is also increased to keep the exit pupil matched to the iris. However, more detail will be visible. When viewing such diffuse objects, the magnification must be chosen so that the exit pupil of the telescope is 8 mm in diameter in order to have the brightest possible image.

The resolving power of a telescope—its ability to form separate images of stars that are close together—depends on a number of factors. Usually the lens of a visual telescope can be made with sufficient perfection that it is not a limitation. Aberrations, too, are

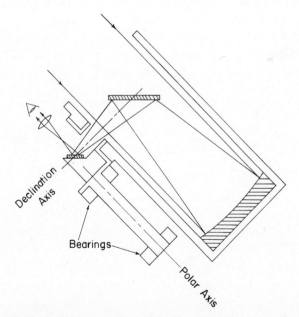

Fig. 49.   The Springfield mounting. This instrument has its eyepiece in a fixed position for convenient observing.

generally not the limitation. The limit frequently is the turbulence of the earth's atmosphere, which results in the shimmering or dancing of star images when seen with a telescope and their scintillation when seen by the unaided eye. Occasionally the atmosphere is steady enough that diffraction and not seeing provides the limitation. For fleeting glimpses this may even be the case for telescope apertures up to 40 inches. When the air is steady, a star's image as seen through a telescope consists of a central disk surrounded by fainter rings. Two stars may be said to be resolved when the disks of the two star images just touch at their edges. An empirical rule has been found to relate the diameter of the telescope to its resolving power. When the diameter of the telescope is D inches two stars $4.56/D$ seconds apart may be resolved. When the two components of a double star are too close to be resolved, the star can often be recognized as double by the elliptical shape of the image. From the amount of the ellipticity an estimate can be made of the separation.

Astronomers have devised a scale for the disturbance caused by seeing. The seeing is usually rated on a scale of 0 to 10, zero being the poorest seeing likely to occur and 10 corresponding to seeing that causes no reduction in the image quality. In practice, since seeing is subjectively determined, the number describing the seeing condition varies according to the size of the telescope used, the magnification employed, and the astronomer making the estimate. There has been some attempt to standardize the seeing scales. Table 3 gives a scale that is frequently used. The diameter of the seeing disk is

TABLE 3.   SCALE FOR RATING SEEING.

Scale number	Diameter of seeing disk (seconds of arc)
0	10.0
1	5.0
2	2.5
3	1.25
4	0.6
5	0.3
6	0.15
7	0.08
8	0.04
9	0.02
10	0.01

taken to be either the average excursion observed for the diffraction disk if this can be observed (*hard seeing*) or the diameter of the diffuse disk that is observed (*soft seeing*). Seeing 2 occurs most frequently for many observatories, while seeing 6 and above occurs very infrequently.

Comparison of Refractors and Reflectors

The refractor is the first choice among most experienced visual observers. Exactly why this should be is difficult to say, since one would expect that the reflector, with its complete freedom from chromatic aberration, would give much superior images. Both types of objective can be made sufficiently accurate to produce equally good images when tested on an optical test bench. The surface accuracy required for equal performance is different in the two cases. The reflector has only one surface compared to the refractor's four surfaces. But, in order to have the same quality of image, the mirror surface must be made to about four times the accuracy of the lens surfaces.

The performance in the observatory is apt to be considerably different for the two telescopes. The difficulty comes from the continually changing temperature inside the dome during the night. The mirror is markedly affected by this change in temperature, for its two sides usually are not equally exposed to the surrounding air and so they change temperature at different rates. This causes the mirror to warp out of shape. A large reflecting telescope will sometimes present multiple images as a result of this warping. The effect can be reduced considerably by constructing the mirror of Pyrex glass rather than of plate glass, which was used for the older mirrors. Pyrex has only one-third the expansion coefficient of plate glass. The thermal effects can be nearly eliminated by the use of fused quartz, but disks of this material cannot be obtained in large sizes.

A lens is also warped by a change in temperature but distortions so introduced at the front surface are largely nullified by similar distortions at the rear surface. Refractors are consequently much less subject to temperature changes than are reflectors.

Diffraction of light by the spider supporting the secondary mirror is a frequent complaint, and the spikes that are produced about star images have been mentioned. The evil effects of this diffraction can be eliminated only by eliminating the spider. This may be done by using the Herschel form of telescope or by supporting the diagonal from an optical glass plate instead of the spider. In the latter arrangement there is still objectionable diffraction caused by the secondary mirror. Its effect is to increase the intensity of the diffraction rings above those presented by the unobstructed aperture. The brighter diffraction rings reduce contrast in planetary images.

Another argument in favor of the refractor concerns its closed tube. Some of the seeing effects found at observatories may actually arise within the dome. Sources of heat, such as motors and humans, produce turbulence of the air. The turbulence cannot enter the closed tube of the refractor. Thus, in spite of the chromatic aberration of the refracting telescope, much can be said in its favor. If color filters are used to eliminate its main disadvantage, it certainly is the superior telescope for visual work.

Requirements of Modern Instruments

The application of the photographic plate, following the experiments of George P. Bond in 1850, revolutionized observational astronomy. The plate could, in a properly designed camera, record large areas of the sky at once, or it could in long-focal-length instruments record star positions far more accurately and quickly than could be done by visual measurements. Along with the advent of photography came a host of other types of observing, such as spectroscopy, photoelectric photometry, and other innovations, which have their special needs and consequently demand the most in versatility in large telescopes.

The long-focal-length refractors, designed for visual observing, were not suitable for this new astronomy. The instruments that are required have to be able to give sharp images over a tremendously greater range of wavelengths than previously. The infrared and ultraviolet ends of the spectrum are frequently not eliminated as the visibility curve of the human eye removes them in visual observing. In fact, the photographic plate is most sensitive in the violet and ultraviolet spectrum and the sensitivity of all plates extends at least to the blue. The shift in wavelength range means that the wavelength of achromatism of refractors has to be shifted to the blue part of the spectrum or, better still, a completely achromatic reflecting telescope should be used. Where plates sensitized to the red are employed, the achromatized range has to be extended by use of more kinds of glass and consequently more lens elements or by resort to mirrors.

Until the 1930's the part of the ultraviolet that could be observed was not limited by the photographic process, nor by the atmosphere, but by the decline of telescope efficiency due to either the poor reflection of mirrors or the nontransparency of glass in the ultraviolet. Either of these effects cut off the ultraviolet spectrum at about 3500 A. Astronomers, having learned much by the extension of 500 A brought about by photography, wanted the remaining 500 A of that still available before the atmosphere made the final limitation at 3000 A. This extension finally came about largely from the development of new reflecting coatings for telescope mirrors.

Reflectivity of Mirrors and Transmissivity of Lenses

Light is lost in a lens in two ways. One way is that the glass itself absorbs some of the light, more so in the short-wavelength region than elsewhere. Flint glass, required for achromatizing a lens, does not transmit below 3600 A and has reduced transmission below 4200 A. In the thicknesses demanded by lenses of large diameter the absorption in even the visible spectrum is appreciable. Figure 50 shows the transmissivity of the Lick 36-inch refractor when used with its photographic correcting lens, which shifts the achromatized region to the blue. Glasses have been developed which transmit further into the ultraviolet but these restrict the optical designer because they do not have the range of dispersion and refractive index desired.

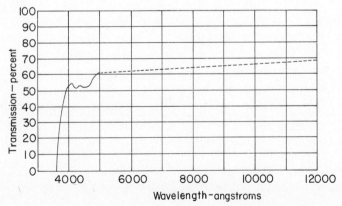

Fig. 50.   The transmissivity of the Lick 36-inch refractor.

The second way in which light is lost in a refractor is by reflection at the surfaces of the lenses. Although small-diameter doublet lenses may be cemented at their interfaces to reduce this loss, large lenses cannot be cemented because of the attendant strains that are introduced as the temperature changes. The light lost at each surface then becomes the fraction $(n - 1)^2/(n + 1)^2$ of the light entering the surface. For crown glass this fraction is about 4 percent, while for flint glass it is about 5 percent. For a doublet refractor with four uncemented surfaces the light lost by reflection amounts to 17 percent. The light lost in this way by a three-element lens is about 25 percent. What may sometimes be worse than the actual loss of light in star images is that some of the light is diffused over the entire plate. Also, a bright star may produce a halo on some other part of the plate than where the primary image appears. The diffused light and the halos obscure and hide detail. The loss of light by reflection, the resulting diffused light, and the halos can all be reduced by the process of coating the glass surfaces with a layer of material of low refractive index whose thickness is one-quarter of the wavelength of the light chiefly to be used. The process for applying this coating is described in the next section.

Modern mirrors are made of glass covered with a thin metal film, usually silver or aluminum, to give a highly reflecting surface. When the metal film has tarnished or otherwise deteriorated, it may be readily renewed without affecting the accuracy of the glass surface.

The reflection coefficients of mirror coatings depend on the angle of incidence and on the polarization, the reflection of light polarized parallel to the mirror surface being different from that of light polarized in the plane of incidence. The reflectances for silver and aluminum are shown in Fig. 51 for different angles of incidence and for two directions of polarization. The figure shows that mirrors used at an angle of incidence of 45° (such as the diagonal mirror in the Newtonian telescope) introduce polarization which makes measurements of the polarization of star light impossible without compensation. Not only is polarization introduced but any linear polarization of the star light is modified. There is a phase shift of the light upon reflection from a metallic surface and it is different for the two polarizations. Incident light which is polarized along a direction not coincident with or perpendicular to the plane of incidence will become elliptically polarized. Fortunately, the effects of the mirror may

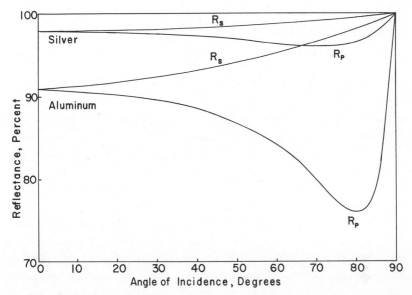

Fig. 51.   The reflectances of aluminum and silver for different angles of incidence, when the plane of polarization is parallel (R_p) and perpendicular (R_s) to the plane of incidence.

be compensated. One way in which this may be done is by the addition of another mirror which bends the light at 90° again but this time at right angles to the plane of incidence of the diagonal mirror. To insure compensation by this means, the second mirror must have exactly the same treatment as the first since the effects produced depend to an appreciable degree on grease films, age, coating conditions, and so forth. The consequence is that both mirrors should be coated at the same time, cleaned at the same time, and be similarly protected. In a coudé telescope, the angle of incidence on the diagonal mirror is not constant. Therefore different compensation has to be applied at different angles of incidence. In practice a mica plate and inclined glass plates can be used to remove any polarizing effects of the diagonal mirror for a small range of declination angles in a coudé. A set of compensating plates is necessary to provide compensation at all declination angles. However, these polarizing properties have to be compensated only in making polarization measurements or spectroscopic studies of the Zeeman effect in stars with magnetic fields in their atmospheres.

Aluminizing and Coating

Before the discovery of the process of silvering glass, about 1850, mirrors were made of speculum metal, an alloy of about 68 percent copper and 32 percent tin. It is very hard and brittle and will shatter if dropped. As a result of its hardness and brittleness it may be ground and polished by the same techniques used for glass, which was the main reason for its adoption for mirrors. Unfortunately, it reflects only about 60 percent of the light that falls on it even when freshly polished, and also it tarnishes to a much lower brilliance than when it is new. After a while, it must be repolished and in doing this care must be exercised to retain its optical figure.

The introduction of silvering on glass and the emphasis on photography made the reflector telescope preferred by professionals. Chemically deposited silver has several disadvantages, however. It may tarnish appreciably in a matter of weeks, it has to be burnished after being coated, resulting in scratches which scatter light, and it has poor reflectivity for ultraviolet light, being useless for wavelengths less than 3300 A. The deposition of aluminum by evaporation in a vacuum was developed for astronomical mirrors by John Strong about 1931. Aluminum is now used on nearly all professional reflectors; it has high reflectivity in the ultraviolet as well as the visible, it scatters relatively little, and a coating that is protected by a cover when not in use will last almost indefinitely.

The vacuum for the deposition of the aluminum must be sufficiently good that the aluminum atoms can travel in straight lines from the place where they are evaporated to the mirror surface without suffering collisions with either gas molecules or other atoms of aluminum. The maximum gas pressure that can be allowed depends on the length of path, but it usually must be less than 10^{-4} mm-of-mercury. If the vacuum is not adequate, the aluminum partly combines with gas molecules in the path and the resulting coating is bluish or grayish and does not stick tightly to the glass surface.

The production of high vacuums in chambers large enough to coat the largest telescope mirrors was once a considerable engineering feat. Today it is done easily by immense diffusion pumps which have pumping capacities of 300,000 liters per minute or more. In such a pump a special oil is boiled under partial vacuum conditions to provide a swiftly moving vapor. This vapor is directed from jets at the center of the throat of the pump barrel to flow along the barrel. As it does so, expanding into the throat, it bumps into gas molecules that enter the throat from the vacuum chamber and knocks them along to the other end of the barrel where they are pumped out by a mechanical pump. The oil vapor is condensed and led back to the boiler, where it is used again.

A schematic diagram of the equipment needed for coating mirrors is shown in Fig. 52. An oil diffusion pump having a high speed (with a throat diameter of 6 to 24 inches, depending on the size of the tank to be evacuated) performs the evacuation to the low pressure. This pump has to be backed up by a mechanical pump capable of maintaining the pressure below the maximum "backing" or "fore" pressure of the diffusion pump (usually about 0.1 mm-of-mercury). The mechanical pump exhausts to the atmosphere the gas that it receives from the diffusion pump. However, most of the gas in the tank must be removed before the diffusion pump can be operated. This is done with a roughing pump similar to the backing pump. Frequently these two are the same pump, which can be connected either directly to the tank through a bypass valve or to the output of the diffusion pump. It is convenient if a high-vacuum valve is introduced between the diffusion pump and the tank. If this valve is closed, the chamber may be opened immediately after the evaporation is completed without having to wait for the pump oil to cool sufficiently that it will not oxidize.

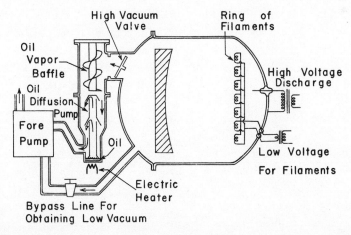

Fig. 52.   The apparatus required to evaporate aluminum onto a large telescope mirror.

The tank is usually made of steel and must be free of even the tiniest leaks. The materials that are put inside the tank must have low vapor pressures so that they do not result in sources of gas, which prevent obtaining a high vacuum. Such sources of gas are called virtual leaks.

Bits of aluminum wire are wrapped around the loops of tungsten filaments arranged in a ring or rings so as to provide a uniform deposit of the aluminum over the face of the mirror. The filaments are heated by a heavy electric current which melts the aluminum, and since molten aluminum wets tungsten and also has a high surface tension, a droplet of it will cling to the filament during the evaporation. Sometimes, however, a drop of molten aluminum will fall and for this reason it is better not to have the mirror underneath the filament array, but rather the mirror should be vertical and the array parallel to it, as in Fig. 52.

The evaporated aluminum atoms travel at high velocities to the surface of the mirror and the walls of the tank, sticking where they strike, and thus building up a coating of metal.

Recent research has shown that better ultraviolet-reflecting films are obtained if the evaporation is made in less than 30 seconds. For this reason all the filaments should be heated simultaneously. The purest obtainable aluminum should be used for telescope mirrors and should not have any protective coating. The highest reflectivity over an extended wavelength range with a relatively long mirror life is obtained in this way. Use of pure aluminum prevents electrolysis, which may occur within an alloy coating, resulting in deterioration.

Cleaning the mirror surface is difficult because all traces of foreign matter must be eliminated before the mirror is placed in the vacuum chamber. The authors' experience is that strong detergents and clean cotton will work about as well as anything for this. After thorough scrubbing, the mirror is carefully rinsed with water (distilled water is preferred though not often necessary) and then dried with clean cotton. Extreme care should be taken at this point that the hands do not touch the surface. Any traces of grease may be discovered by breathing on the mirror. If a uniform dark film forms, one may be sure the surface is clean. During the pump-down, when the pressure is between 1 mm and 10^{-3} mm-of-mercury, a high-voltage glow discharge should be produced within the chamber. The discharge burns off any residual grease remaining on the mirror surface.

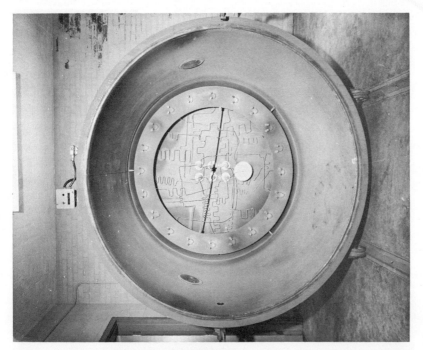

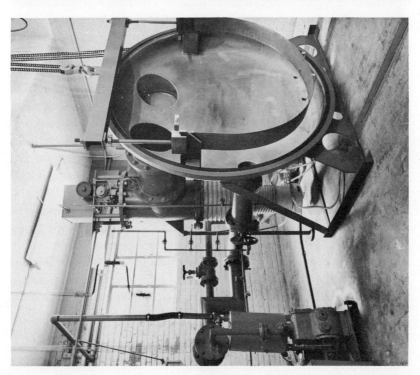

Fig. 53.   The aluminizing equipment for the 61-inch telescope of the Harvard College Observatory.

To facilitate the burning a small amount of oxygen is introduced.

Figure 53 shows the aluminizing tank and evacuating equipment at the Harvard Observatory for coating the 61-inch mirror. The tank is open; it rolls back away from the base frame and pumping equipment on tracks. Two hooks, which may be seen attached to the base frame, support the mirror, which is slung in a stainless-steel band. A carriage, not shown, transports the mirror from the telescope tube to the tank and safely tilts it from the horizontal position to the vertical. From here the overhead crane is used to remove the mirror and the sling and to place the sling on the hooks of the base frame. The inside diameter of the tank is 72 inches, allowing ample room for clearance of the mirror and the sling. A single 16-inch-throat diffusion pump together with the roughing pump, with a capacity of 50 cubic feet per minute, evacuates the tank in a little over 2 hours.

Inside the tank are 20 filaments each consisting of five turns, ½ inch in diameter, of three-strand tungsten wire. The filaments are arranged in a ring of diameter 50 inches. The ring is 30 inches from the mirror surface.

Lenses are coated to reduce surface reflection in a manner very similar to that first used by Strong. The material for coating is either magnesium fluoride or cryolite. These crystalline materials cannot be hung from a filament and consequently must be evaporated from a boat—a folded strip of tungsten or other refractory metal that can hold a liquid and be heated by electrical conduction. The evaporation must therefore take place upward and the lens to be coated must be positioned above the boat. A harder coating can be obtained if the glass is heated by radiation from a heating coil in the tank to a temperature of several hundred degrees centigrade before the coating. This process, however, is dangerous except for small lenses. The depth of the coating must be precisely controlled to provide a film that reduces the reflectivity of the glass surface to a minimum for the desired wavelength. The thickness of the coating may be checked in several ways. Its color may be watched and the evaporation stopped when it has attained the proper color, a dark purple if the minimum of reflectivity is to be for green light. A photocell and light source may be enclosed in the chamber and arranged so that the surface to be coated reflects the light into the photocell. The process is stopped when the photocell reaches the proper reading (determined by experiment).

Angular Field, Plate Scale, and Resolution of Telescopes

The introduction of the photographic plate emphasized other requirements of telescopes besides those of achromatism and ultra-violet transmission. The human eye is satisfied to view only a small region of the sky clearly at any one time. If a nearby area is to be viewed, the telescope can easily be moved to this area. The photographic plate, however, is not limited by anything similar to the fovea of the eye; it can record, at the same time, sharp images over its entire surface. Thus, telescopes that can produce sharp images over the whole area of the plate are much desired. Inherently the photographic telescope records over a much larger region of the sky than an observer with a visual instrument can hope to cover.

The actual field to be imaged in a telescope depends chiefly on the problem to be studied. Telescopes whose chief purpose is to survey large stellar fields prior to more detailed work by other instruments, or to obtain information through investigations on a large number of objects, have to be wide-angle instruments. The definition required is not as high as that necessary for some other purposes but it should be uniform over the field. Telescopes for the study of the distant extragalactic nebulae need only a moderate field but require large light-gathering power. Similarly, instruments that are intended for stellar spectroscopy and for stellar photometry require at most a field sufficient for finding the object and guiding on it. For photography of planets and double stars the highest resolution is required, but because of the restricted size of the object the necessary field is very small. To try to combine the highest resolution with a very large angular field in one instrument is out of the question. The objects to be investigated do not require it and the size of the plate needed would be too immense to be convenient.

An important characteristic of photographic telescopes is the *plate scale*, which is the number of seconds of arc per millimeter along the plate. Cameras intended for photography of the Milky Way may have scales of several thousand seconds per millimeter, while telescopes designed for planetary photography may have a scale of from less than one to a few seconds per millimeter. A large plate scale, that is, a small number of seconds per millimeter, means a long focal length for the objective. The plate scale is connected to the focal length by the relation $s = 206.265/f$, where s is the plate scale in seconds of arc per millimeter and f is the focal length of the objective in meters.

The angular fields covered by astronomical cameras vary from 140° in an instrument designed to photograph the Milky Way, the aurora borealis, and the zodiacal light to only a few minutes of arc required for planetary or double-star photography. The more usual fields that are covered range from 6° or 7° in Schmidt cameras to 0.5° to 0.25° at the Newtonian focus of a large reflector.

Numerous factors limit the resolution of a photographic telescope. Ideally, a star image formed by a telescope would be extremely small. (The star with the largest apparent diameter, Antares, subtends an arc of only 0.065 second, and most stars have apparent diameters far below this.) Actually, the image on the photographic plate has a quite noticeable size. Some of the limiting factors, including diffraction and seeing, have been discussed under visual observing. Diffraction results in an otherwise perfectly imaged star having a diameter of $4.56/D$ seconds of arc, where D is the diameter of the telescope in inches. Selection of the observatory site can improve seeing. Some of the best seeing has been obtained at observatories in the western part of the United States, but no doubt sites giving equally good seeing are obtainable in almost every part of the world. Turbidity of the photographic emulsion is a resolution-limiting factor, discussed in Chapter 2. The light from a star image is scattered and diffused in the emulsion. For an average emulsion and normal exposure the diameter of an otherwise point star image is about 0.02 mm. Unsteadiness in the pointing of the telescope during the exposure is another cause of poor resolution. The reason may lie in wind blowing the instrument, failure of the driving clock to run smoothly, or inadequate guiding of the telescope during the exposure. A further cause of the finite size of stellar images may be poor quality of either the optical surfaces of the telescope or the corrections built into the original design. The latter effect is particularly important in instruments designed to cover a wide field, since the angular coverage of a given design is pushed to the point at which those errors that depend on the angular extent of the field become important.

We have already discussed the two types of aberration, chromatic and spherical. When star images not on the axis of the camera are considered, three other types of aberration come into prominence.

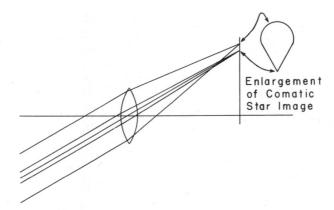

Enlargement
of Comatic
Star Image

Fig. 54.   The formation of comatic images by a lens.

The first of these, *coma,* is an aberration that is generally important for star images that are off the axis of the telescope by only a small angle. The name arises from the characteristic cometlike appearance of the star images. The aberration results from the difference in magnification produced by different zones of a lens or mirror. In Fig. 54, it is shown that the central zone of a simple lens magnifies less than the outer zone. The appearance of the comatic pattern is sketched on the right in the figure; the sharp point of the triangle is the brightest part of the pattern. For a given lens, the extent of the comatic image increases proportionally to the distance of the star image from the center of the plate. The aberration at 10° from the axis is only twice that at 5° from the axis. However, for different lenses, coma at a given angle from the axis is proportional to the square of the diameter of the lens.

At angles farther from the axis of the telescope the image of a star is frequently dominated by another aberration, *astigmatism.* This aberration results in a star image appearing as a short line, pointing either toward or at right angles to the axis of the lens, depending on how the camera is focused. Formation of an astigmatic image is diagramed in Fig. 55. Consider the plane *AOP* through the axis of the lens and the image point *P*. Rays striking the lens in this plane will be brought to a focus closer to the lens than rays that strike the lens in the plane at right angles to this, that is, in the plane *BOP*. These rays are focused at the point *P'*. The focus *P,* which is closer to the lens, is called the tangential focus, while the other focus *P'* is

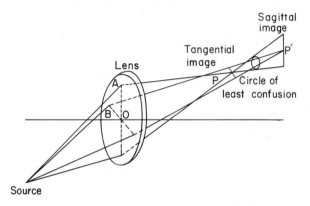

Fig. 55.   The formation of astigmatic images by a lens.

called the sagittal focus. When all the rays from the lens are considered, the image at *P* is a short line tangent to a circle about the
axis of the lens, while the image at *P'* is a line pointing toward the
axis of the lens. The best focus is obtained between these two positions, where the circle of least confusion is obtained.

Astigmatism increases as the square of the distance of the star
image from the axis of the lens. Thus the astigmatism at 10° from
the axis of the lens will be four times that at 5°, so that astigmatism
will dominate the coma at large angles from the axis of a lens. On
the other hand, coma will be the larger at small angles from the axis.
For different lenses, astigmatism increases only proportionally to the
diameter of the lens.

The third aberration of importance to images not on the axis of
the optical system is *curvature of field*. In this aberration the surface
where the best focus is obtained is not plane but has a more or less
spherical shape. If the curvature is not too large, a glass photographic
plate may be bent to fit this shape and then the aberration does not
result in poor images. A plate that is to be given a convex shape
can be readily bent over a steel form having the proper curvature,
the plate being retained at the edges. A plate that is to have a concave curvature may be bent by evacuating the space between it and
a steel form of the proper curvature, so that atmospheric pressure
forces the plate against the form. Thin glass of the order of 1 mm
thick should be used for this purpose. The amount that a plate can
be bent is surprising; in the 48-inch Schmidt camera, a 14-inch
square plate is bent to a radius of 10 feet. If the plate cannot be so

bent, then parts of it necessarily have to be out of focus and the size of stellar images is increased at such parts.

There is a fourth off-axis aberration, *distortion,* which does not increase the size of stellar images but only results in different magnification or plate scale in different parts of the field of the camera. In cameras for astronomical use it is more of a nuisance than a serious disadvantage.

The highest possible resolution is desired for planetary photography. Here the major limitations on the resolution will be the atmospheric turbulence and the turbidity of the plate emulsion. This second limitation comes about from making a compromise to gain some advantage on the seeing. A dancing star image may be stationary occasionally for a fifth to a tenth of a second, and very rarely for as much as a second. If a long series of photographs is taken with short exposure times there is a chance that a few of them may have been taken during these moments of good seeing. In order to keep the exposure times sufficiently short, a fast photographic emulsion is required, and such emulsions usually have lower resolution than slower ones. Planetary photographers have differing opinions on this subject, however, and some make this compromise while others do not.

Telescopes, as a rule, do not have the long focal lengths built into them that are required for planetary photography. The large plate scale is obtained by an auxiliary enlarging lens. Either a positive or a negative lens may be used. The negative lens is placed in the converging beam from the objective and makes the effective focal length considerably longer. The positive lens is placed after the focus and reimages it with several times magnification. Such lenses for visual refractors may be designed to shift the achromatism of the telescope to the color in the spectrum in which it is desired to make the photographs. Usually two enlarging lenses will suffice, one for photography with blue light and one for the yellow, red, and infrared.

The choice of reflector or refractor for planetary photography is also a question in dispute. Many prefer the coudé focus of a large reflector to take advantage of the large amount of light and consequently shortened exposure time afforded by such instruments. Others prefer to use a refractor, and may even place a stop in front of the lens to limit the aperture, depending on seeing conditions. Both reflectors and refractors have taken very excellent planetary photographs.

Large Reflectors

The work horse of much present-day observational astronomy is the large reflecting telescope. It is generally made to be several telescopes in one—Newtonian or prime focus, Cassegrainian, and coudé—because such a telescope is an expensive instrument and must be as useful and versatile as possible. Its primary uses are direct photography at the prime or Newtonian focus, spectroscopy of different kinds at all of the foci, and photoelectric photometry at either the Newtonian or the Cassegrainian focus. The largest reflector is the 200-inch Hale telescope on Palomar Mountain (Fig. 85). A photograph it has made of an external galaxy is shown in Fig. 56.

Because of the small field that is covered in a large reflector and because of the perfect correction of spherical aberration and the non-existence of chromatic aberration, the only aberration of importance

Fig. 56.   The spiral galaxy NGC 1300 photographed with the 200-inch. (Mount Wilson and Palomar Observatories photograph.)

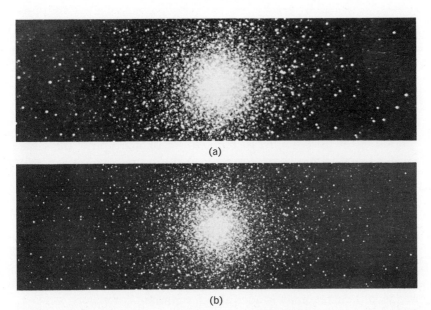

Fig. 57. Correction of coma by the Ross coma correcting lens: (a) photograph of the globular cluster M 3 without the lens; (b) photograph of the same cluster with the lens. (Mount Wilson and Palomar Observatories photograph.)

is coma. As larger telescopes were constructed, this aberration became more important. If the f-number is kept constant as the size of the reflector is increased, the focal length becomes proportionally larger. It is then desirable to employ larger plates to maintain the same field. But as the focal length is increased the comal aberration is also increased proportionally and becomes intolerable at the edge of the plate. If the plate size remains fixed, instead of increasing, the coma at the edge of the plate remains constant, since the coma is proportional to the field angle. Thus the image errors are no worse in terms of linear size at the edge of a 4 × 5-inch plate on a 40-inch f/5 telescope than at the edge of the same plate on a 120-inch f/5 telescope. However, it is desirable to keep a large field and a not too prohibitive instrument size and plate size by using a smaller f-number, say f/3. At this value, coma becomes intolerable at just a short distance from the axis.

Figure 57(a) is a photograph of a star cluster taken at the f/3.3 prime focus of the 200-inch reflector. The coma becomes noticeable at just 1 mm from the axis of the telescope, and without some means of correcting it the 200-inch would have been almost useless for photography at the prime focus. Fortunately, the coma of a reflector may be corrected by the *zero corrector,* a combination of two lenses that are slightly larger than the plate size and placed just in front of the focus. Although it is called the zero corrector because it usually does not change the effective focal length of the telescope, that is, has zero power, in the 200-inch the corrector was designed to convert the $f/3.3$ ratio of the primary mirror to $f/3.6$ for better correction of the aberrations. In Fig. 57(*b*) a photograph of the same cluster taken with a coma corrector at the prime focus of the 200-inch shows the effectiveness of the correction; this sharp photograph would not have been possible without it.

The guiding of a plate is performed in either of two ways: the plate is arranged to be moved in two coordinates by two screws, or the telescope may be moved by slow motions that are adequately slow and without backlash or overtravel. An eyepiece with cross hairs is fixed to view a star at the edge of the field of the plate. This image, although comal, is adequate for guiding.

Spectroscopic observations are made at either the Newtonian focus or the coudé and Cassegrain foci, depending on the dispersion required. A high-dispersion spectrograph requires a large temperature-controlled room and so the coudé focus is frequently used. Intermediate-dispersion spectroscopy can be performed at the Cassegrain focus. The low dispersions demanded by faint nebulae are best obtained with efficient spectrographs attached to the Newtonian or prime foci.

The focus used for photoelectric photometry is a matter of choice. The long focal lengths afforded at the Cassegrain focus offer advantages because of the large scale.

Astrographic Cameras

Astronomical cameras that can image an area of the sky 5° or more in diameter at a moderate plate scale are required for survey work and accurate recording of star positions. As has already been explained, a relatively large *f*-number is desired for imaging the faintest stars, while for imaging extended nebulosity a small *f*-number is better. A number of different cameras using refracting or reflecting optics or a combination of both are available. We will discuss the *dioptric* or refracting cameras first.

The Astrographic Doublet. One of the first lenses applied to photographic astronomy was the Petzval portrait lens. It consists of two positive achromatic doublets and it is corrected for major aberrations. A modification of this lens, the astrographic doublet, has a negative achromat for the second compound lens. With this change, nearly complete correction for curvature of field is possible. A further advantage of this change is that a telephoto effect occurs, that is, the second element magnifies the focal length of the primary (as does the secondary mirror in a Cassegrain telescope) and the resulting focal length is longer than the total length of the camera. The lens can be made with speeds up to $f/7$ and can cover a 10° field.

The Cooke Triplet. This lens was designed by H. Dennis Taylor of the firm of T. Cooke and Sons at the turn of the last century and it represented a new departure in lens design. Taylor used only three simple separated lenses (Fig. 58) and found that he could flatten the field and correct astigmatism at the same time by making one of the lenses negative and of a power equal to the sum of the powers of the positive lenses. Because the lenses are separated, however, the combination could have positive power and correction for all important aberrations. It can have an *f*-number of $f/5$ for short focal lengths and $f/7$ for longer focal lengths. The scheme of design used in the Cooke triplet is adapted for more complicated lenses, such as the Zeiss Tessar; in this lens, the last element is a compound, positive and negative lens. The Cooke lens has to be kept in very accurate adjustment. This requirement for a very rigid mount is not unusual for a photographic lens and much attention should be paid to design of the mount to maintain the adjustment.

The Ross Lens. This well-known lens of four elements (Fig. 59) is widely used in astronomy. It can give a highly corrected field of 20° at an *f*-number of $f/7$ or even less. The largest lens of this type has an aperture of 20 inches and a focal length of 144 inches. It is used

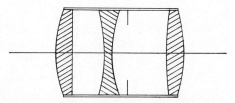

Fig. 58.　The Cooke triplet camera lens, $f/5$.

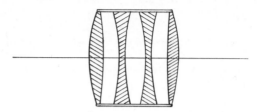

Fig. 59.   The Ross astrographic lens, $f/7$.

at the Lick Observatory in a program for stellar proper-motion study. The Ross lens can be corrected to give sharply defined images in one color. Although it is corrected for chromatic aberration, the residual color error is large and the corrections for spherical aberration depend markedly on the wavelength. At one wavelength the spherical aberration is satisfactorily corrected but at other wavelengths the correction is poor. The result is a hybrid aberration called chromatic difference of spherical aberration. The Ross lens also suffers from chromatic difference of astigmatism, which is a similar hybrid aberration where the astigmatic correction fails for some wavelengths.

Lenses Adopted from Photography.   A number of lens designs originally made for hand cameras have been used for astronomical purposes. One of these is the Zeiss Biotar (Fig. 60). It can be used with f-numbers as low as $f/1.4$ and angular fields of $30°$. It is recommended only for short-focal-length survey work, however, and functions considerably better at $f/2.5$. Many of the photographic lenses that have been applied to astronomical uses have poor images, a large amount of the light going into halos around the stellar images. Though such spreading of light in halos is not too serious for pictorial photography, in astronomical applications it may completely obscure faint detail.

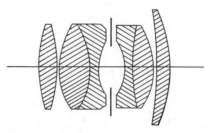

Fig. 60.   The Zeiss Biotar lens, $f/1.4$.

Mirror Cameras

Several instruments having two or more mirrors, which give a wider field than the parabolic reflector, have been developed. Systems having two curved mirror surfaces can be designed that are free from spherical and comatic aberration. Unfortunately, the primary mirror of such systems is not parabolic so that it cannot be used alone with a Newtonian focus.

Schwarzschild Reflector. The German astronomer, K. Schwarzschild, investigated all possible combinations of two mirrors that were free from spherical aberration and coma. The best-known of these (Fig. 61) has his name attached to it. Besides being free from spherical aberration and coma, it has the added advantage that it has no curvature of field. On the other hand, the large secondary mirror obscures about one-fourth of the light that would otherwise fall on the primary. A more serious disadvantage is that the two concave mirrors and the light baffle, needed to prevent sky light from falling directly on the plate, necessitate a long tube and consequently a large dome. The largest instrument of this type, at Indiana University, has a 24-inch aperture and an 80-inch focal length.

Ritchey-Chrétien Telescope. The correction for coma in this instrument (Fig. 62) was discovered empirically by Ritchey, who found that a slightly oblate spheroidal primary mirror when combined with an elliptical convex secondary was free of coma. However, it

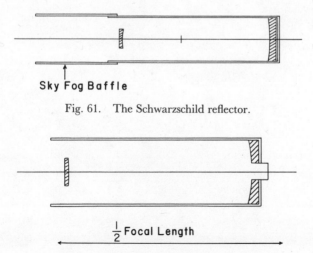

Sky Fog Baffle

Fig. 61.   The Schwarzschild reflector.

$\frac{1}{2}$ Focal Length

Fig. 62.   The Ritchey-Chrétien telescope.

does require a curved plate for best results. It has the advantage that the tube length is considerably less than the focal length. It is recommended for *f*-numbers no less than *f*/6. The largest instrument of this type (Fig. 63) is at the U. S. Naval Observatory at Flagstaff, Arizona, and has an aperture of 40 inches.

Fig. 63.   The 40-inch Ritchey-Chrétien telescope at the U. S. Naval Observatory at Flagstaff, Arizona. (Official U. S. Navy photograph.)

The Schmidt Camera

It is surprising that most of the remarkable systems that can result when lenses are combined with mirrors were not discovered until after 1930 when Bernhard Schmidt discovered a very useful property of the spherical mirror when combined with an aspherical correcting lens. Before delving into *catadioptric* systems, as optical instruments which combine lenses and mirrors are called, let us investigate the image-forming properties of the spherical mirror with a stop at its center of curvature.

An image formed by a spherical concave mirror of an object at infinity has a large amount of spherical aberration similar to that illustrated in Fig. 35 for a lens. Rays intercepted by the mirror in a zone near its rim focus closer to the mirror than those intercepted near its center. Figure 64 shows a spherical concave mirror with an aperture or stop placed at its center of curvature. Light from a star coming in perpendicular to the stop is focused into a poorly formed blob at a point midway between the center of curvature and the mirror surface. Now consider the light coming from a star at a considerable angle to the light from the other star. The central ray of this beam of light must also pass through the center of curvature of the mirror and so the alignment of center of curvature, focus, and mirror surface is exactly the same for the second beam as it was for the first. The image that is formed then has exactly the same amount of aberration that the first image had. The system having just a spherical mirror with the stop at the center of curvature has the re-

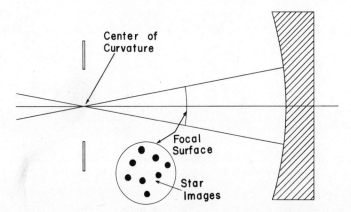

Fig. 64.   A spherical mirror with a stop at its center of curvature produces uniform images over its entire field but with a large amount of spherical aberration.

markable property that it is free from chromatic aberration, coma, astigmatism, and distortion, and it suffers only from spherical aberration and curvature of field. If the spherical aberration of this system could be lessened or removed without introducing too much of the other aberrations, the image quality would be excellent. The remarkable thing is that the spherical aberration can be removed by the addition of one lens; the curvature of field can be eliminated by the addition of another.

In 1930, Bernhard Schmidt, an optician at the Hamburg Observatory, discovered one way that the spherical aberration could be corrected. He added a thin aspherical correcting lens (Fig. 65) at the center of curvature of the mirror. The lens was nearly plane but was so shaped that it carried just the correction needed for the spherical aberration that the mirror was going to introduce. Schmidt knew that the figuring necessary to eliminate spherical aberration in a conventional reflector did not have to be performed on the primary mirror, for he had often put the correction on the secondary mirror of a Cassegrain telescope. Apparently he reasoned that the correction could be put on a plate of glass instead of on the mirror surface and that the logical place for this plate was at the center of curvature. The aspherical correcting plate was difficult to produce; nonetheless, the camera resulting from its use had excellent properties and justified the work.

Just as a spherical mirror may be parabolized by deepening the center or by grinding the edge or by a combination of the two, the correcting plate may be figured in many different ways. Figure 66 shows the range of shapes that the plate may take. Here the vertical scale is much exaggerated to make the forms apparent; on an actual lens the departure from a plane surface is barely noticeable from a slight distortion of the shape of objects reflected in it.

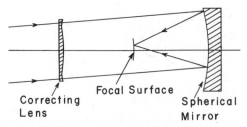

Fig. 65.   The Schmidt telescope. A spherical mirror is combined with a nearly plane correcting lens.

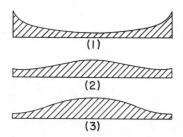

Fig. 66.   Possible shapes of Schmidt correcting lenses. The departures from plane surfaces are much exaggerated.

The image formed by the Schmidt system is excellent although not perfect. The correcting lens has the correct shape for removing spherical aberration for only one wavelength of light. For other wavelengths some aberration remains. Then, too, there is a slight shift of the focus of the system with color, that is, the correcting plate introduces chromatic aberration. Another reason for the formation of imperfect images is that the correcting plate introduces some coma. The lens has the correct shape for removing the spherical aberration from the light that enters perpendicular to the plate. A thin prism deviates a light beam according to the angle of the prism and almost independently of the angle at which the beam entered the prism. Since the aspherical lens at any point is equivalent to a thin prism, the amount of correction that is introduced does not depend much on the angle of the entering light to the corrector. However, the fact that the plate does not perfectly remove the spherical aberration for images some distance from the axis of the camera has the effect of introducing coma. The Schmidt camera, therefore, has a large field which is highly corrected for rather large apertures or small f-numbers, but there is some residual chromatic aberration, chromatic difference of spherical aberration, and coma. In addition, the surface of best focus is that of a sphere. If the radius of curvature of this surface is not too small the plate may be deformed to fit to this surface. In some cameras that use film, the film may be molded by the application of heat to the correct radius of curvature.

Schmidt cameras of short focal lengths may have f-numbers as small as $f/1.0$ and may cover fields as large as $25°$. Still smaller f-numbers can be made by filling in with glass the space between the back surface of the corrector plate and the front of the mirror.

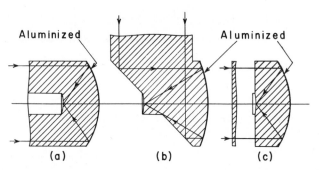

Fig. 67.   Different modifications of the solid-glass Schmidt camera.

This form of the Schmidt [Fig. 67(a)] then becomes one solid cylindrical piece of glass with the correcting surface ground on the front and the spherical mirror ground on the back. The back surface is then aluminized to make it a mirror. The photographic plate is inserted through a hole drilled into the glass from the front of the cylinder. A layer of oil must fill the space between the plate and the glass surface. Such solid Schmidts, as they are called, may be made with f-numbers as small as $f/0.6$. The possibility of such a small f-number results from the fact that the effective focal length, compared to the nonsolid variety, is reduced by $1/n$, where n is the refractive index of the glass. However, the plate is in a very inaccessible location. Variations of the solid camera that do not have the inaccessible plate location are shown in Figs. 67(b) and 67(c). These forms of the Schmidt camera find applications in stellar spectrographs.

In telescopes of long focal length the f-number cannot be made as small as $f/1.0$ and the field covered is much smaller than $25°$. For example, the largest Schmidt camera, on Palomar Mountain (Fig. 84), has a primary mirror 72 inches in diameter and a correcting plate 48 inches in diameter; its f-number is $f/2.5$ and it photographs a field of $7°$. The photographic plates are 14 inches square and are deformed to the 120-inch radius of the focal surface. The plate is loaded by means of a special loading mechanism (Fig. 68). An example of the excellent performance of the 48-inch Schmidt is shown by the photograph in Fig. 69.

The choice of the shape of the correcting plate helps some in reducing the chromatic aberration. The shape shown in Fig. 66, Number 2, has the least chromatic aberration and less coma than other shapes. Another way in which the chromatic aberration may

Fig. 68. Loading the 48-inch Schmidt camera. (Mount Wilson and Palomar Observatories photograph.)

be reduced further is by making a doublet plate of crown and flint glass so that the correction that it affords is without color error. The complication of making two aspheric plates, however, is not often resorted to.

The curvature of field may be eliminated by the use of the field-flattener lens, a device introduced by the astronomer Piazzi Smyth in 1873. This is a plano-convex lens (Fig. 70). The radius of curvature of the convex surface should be $(n - 1)/n$ times the radius of the focal surface that it is to flatten, where n is the refractive index of the lens. In use, the lens is placed nearly in contact with the photographic plate, where it has little effect on the image except to modify the focal surface. However, for f-numbers smaller than $f/2$ it does introduce appreciable coma. A disadvantage of the field

Fig. 69.   Nebulosity in Cygnus photographed with the 48-inch Schmidt camera. The spikes on the bright star images are due to diffraction by supports within the camera. (Mount Wilson and Palomar Observatories photograph.)

flattener is that light is reflected back and forth between its surfaces and the plate and contributes to spurious images and diffused light over the plate.

Modifications of the Schmidt Telescope

Bernhard Schmidt, an imaginative optician, changed the thinking of the professional lens designers. Within ten years the optical designers were working hard to remove the traces of aberrations that remained in his system.

One of the first items to be revised was the correcting lens. In 1940, it occurred independently to two designers, D. D. Maksutov in Russia and A. Bouwers in Holland, that it was not necessary to make an aspherical correcting lens. Rather, a negative meniscus lens with spherical surfaces can be made to introduce spherical aberration

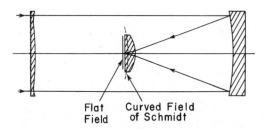

Flat   Curved Field
Field   of Schmidt

Fig. 70.   Use of a plano-convex lens to flatten the field of a Schmidt camera.

without introducing appreciable focusing effect. The fortunate thing is that the spherical aberration it introduces cancels the spherical aberration of the primary mirror. This telescope is now known by the name of Maksutov, though Bouwers has just as much claim to it. Wartime stifling of scientific communication kept Bouwers's contribution hidden until Maksutov's name was firmly attached to it in the United States, as a result of the latter's article in the *Journal of the Optical Society of America* in May 1944.

The telescope that results with a meniscus correcting lens is shown in Fig. 71. By making the lens slightly diverging and choosing the proper radii and thickness, it can be made achromatic. Further, because the lens is slightly diverging, it does not have to be placed nearly so far from the primary mirror to eliminate coma as does the correcting plate in the Schmidt. This results in a considerably shorter tube length and lower telescope cost. A difficulty, however, lies in producing the meniscus lens with its very deep curves. Not only must the blank be molded to the approximate shape before grinding, but in the grinding and polishing operations there is considerable danger of breakage. The optical perfection of the Maksutov relative to the Schmidt depends on the *f*-number of telescope. In general, the Maksutov system is better for the long focal lengths because it can give optically better images than the Schmidt for a given tube length. At very small *f*-numbers, the Schmidt is superior. The

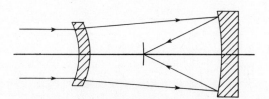

Fig. 71.   The Maksutov-Bouwers camera.

Maksutov design has not solved one of the basic problems of the Schmidt camera, namely, the curvature of field.

One way in which the field can be flattened, by use of the field flattener, has already been mentioned. It would be highly desirable to design the curvature out of the system as a whole. Upon first consideration this may appear impossible, since the curvature comes about as a natural result of the symmetry provided by the spherical mirror. J. G. Baker investigated systems having two reflecting elements and a single correcting lens. Let us first consider the usual Schmidt system with an additional flat mirror as in Fig. 72(a). The plane mirror serves to fold the focus to a position near the spherical primary. In fact, if the primary is provided with a hole, the flat mirror can be made to bring the focus to a point behind the mirror where it is more accessible, a considerable advantage by itself. Usually the auxiliary mirror required to do this will block an appreciable portion of the light.

Now instead of having a flat mirror, suppose we bend the secondary mirror so that it grows gradually convex, all the while moving the correcting plate and changing the primary mirror to keep the field free of spherical aberration, coma, and astigmatism. Finally, the field becomes flat. For each distance of the correcting plate from the secondary mirror there is a corresponding position of the correcting plate that provides a flat field. Unfortunately the required mirror surfaces are not necessarily spherical. Three such systems are illustrated in Figs. 72(b,c,d), where for each system the focal lengths and apertures are the same as for the folded Schmidt system at the top of the figure. Camera b has the property of a very

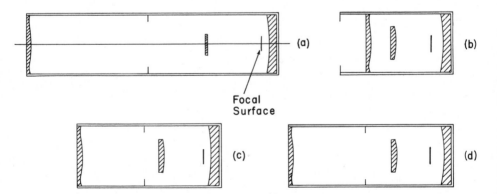

Fig. 72.   The Schmidt camera with plane mirror and its modified forms.

short tube length, but it unfortunately requires both mirrors to be aspherical. Camera *c* requires a spherical primary mirror with a nearly spherical secondary. In fact, the required departure from a sphere for a 12-inch *f*/2.3 system is only ¾ wavelength. This asphericity is not at all difficult for an optician to achieve. Camera *d* is the most practical to build. It will image stars over an 8° field on a flat plate with image diameters smaller than the resolution of the plate.

A telescope built according to this last formula is located at the Boyden Observatory in South Africa. The correcting lens has a clear aperture of 33 inches, while the primary mirror is spherical and has a 36-inch diameter. The total tube length is only 168 inches, compared to 240 inches for the normal Schmidt. The focal length is 119 inches, making the *f*-number *f*/3.6. The circular photographic plate is 10½ inches in diameter. It is 9 inches in front of the primary mirror and can be changed through a large hole in the mirror. The secondary mirror is also spherical and has a diameter of 17 inches. It therefore obscures about 25 percent of the incident light, which is not a serious amount. The field covered by the telescope is 5°.

Other modifications of the two-mirror, correcting-plate telescope have been designed by E. H. Linfoot and by H. Slevogt. One by Linfoot has spherical primary and secondary mirrors and the focal plane lies behind the primary mirror, where it is quite accessible. This system, however, introduces some astigmatism.

Table 4 summarizes the presence of the aberrations in the various telescopic cameras.

TABLE 4.   CHARACTERISTICS OF VARIOUS TELESCOPIC CAMERAS

Camera type	Aberration				Average speed	Tube length
	Spherical	Coma	Astigmatism	Curved field		
Paraboloid	0	*	*	*	*f*/5	medium
Refractor	0	0	*	*	*f*/15	medium
Schwarzschild	0	0	*	0	*f*/3	long
Ritchey-Chrétien	0	0	*	*	*f*/7	short
Schmidt	0	0	0	*	*f*/3	long
Maksutov	0	0	0	*	*f*/3	medium-to-short
Correction-plate, two-mirror	0	0	0	0	*f*/3	medium-to-short

An asterisk (*) indicates an appreciable amount of the aberration.

Still faster cameras with wider fields than those already described have been designed. These combine the correcting plate of the Schmidt with the meniscus lens of the Maksutov. One of these, designed by Linfoot, has a spherical meniscus lens concentric with the spherical mirror and a double aspherical corrector at the center of curvature of the sphere. The meniscus lens removes a large part of the spherical aberration of the mirror and the correcting plate has only to eliminate the remainder.

A camera designed by J. G. Baker has been used extensively for meteor photography. It has two spherical shell lenses (Fig. 73) and a spherical mirror. All spherical surfaces have the same center of curvature. The spherical shells serve to reduce greatly the amount of the spherical aberration produced by the spherical mirror. In addition, since they are centered on the stop and the center of curvature of the primary mirror they introduce no coma or astigmatism. By making the meniscus lenses large enough to include the focal surface as shown in the figure, the second of the two lenses is made to do double duty. That is, on the second traversal of this lens, after being reflected from the primary mirror, the light has additional correction for spherical aberration given to it. The function of the correcting plate (which is made of two aspherical crown and flint lenses, to be achromatic) is to remove the small amount of spherical aberration remaining in the camera. This the weak corrector can

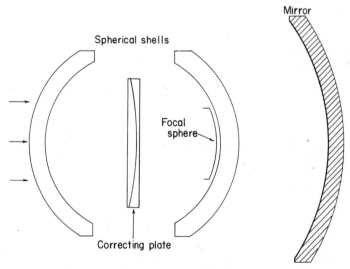

Fig. 73.   The super-Schmidt camera, $f/0.67$.

Fig. 74.   One of the super-Schmidt meteor cameras.

do without introducing appreciable coma or astigmatism. The system can then be designed (for short focal lengths) with the phenomenal combination of an extremely low f-number of $f/0.67$ and a sharply imaged field of $55°$. The effective f-number, however, is only $f/0.85$ because of the obscuration of the film. A number of such meteor cameras (Fig. 74) have been made having a focal length of 8 inches and a 12-inch aperture. The focal surface has a radius of curvature of only 8 inches. Sheet film is molded to conform to this radius at a temperature which is not so high as to damage the emulsion. The film is cut to a circular shape with a diameter of 7 inches across the chord. An exposure time of only 10 minutes with fast blue-sensitive film is sufficient to bring out the sky fog.

Choice of a Telescope

As was mentioned in Chapter 1, there are essentially two types of objects, namely, stars (point images) and nebulae (extended luminous objects), which astronomers are interested in photographing. The choice of telescope for photographing nebulae is treated first.

The difficulty in photographing faint extended objects is that the upper atmosphere of the earth emits very faint light which can be seen only at night. This light of the night sky has recently been named "air glow." In the vicinity of cities there is additional light to contend with and new observatories are now located remote from cities. In the high northern and southern latitudes there are frequently auroras whose light is similar to that of the air glow. Against such backgrounds we have to photograph the object we are seeking. Naturally, very faint objects will be masked by the background light. This sets a limitation on photographing the faintest objects, a limitation that is independent of the telescope used. In photographing the faintest nebulae, then, all that can be done is to expose the plate to bring the sky background to the part of the characteristic curve where the maximum gamma is obtained.

The exposure time required to do this depends on the f-number of the instrument. Neglecting reciprocity failure, the exposure for an $f/10$ telescope will be 25 times the exposure required by an $f/2$ camera. Reciprocity failure will make the required exposure longer still. There is a reason for using the higher f-number, however. For a given aperture size it will have the larger plate scale and so on a fully exposed plate more detail will be seen. If both short exposure times and large plate scale are required, then a large aperture is mandatory. For this reason, a large reflector, such as the 200-inch, is best for photographing distant galaxies (Fig. 56). On the other hand, large extended objects such as the nebulosity in Cygnus (Fig. 69) require a large angular field and a low f-number. Schmidt and similar cameras are then suitable. Presently we shall discover another reason why low f-numbers are desirable for recording nebular objects.

Stellar images present another problem. We have discussed a

number of the factors that establish the diameter of a star image on a photographic plate. Some of the factors are related to the focal length or plate scale, such as aberrations, seeing, guiding errors, and diffraction. Others, including photographic turbidity and halation, do not depend on the plate scale. An important phenomenon was discovered by Hubble: the detectability of very faint star images on plates exposed to show the air glow depends on the image diameter. The larger the images the more surely they may be found. Thus, for telescopes of the same aperture, the one with the longer focal length will record the fainter stars. In addition to the effect of focal length there is the strong effect of the diameter of the objective. If the size of the objective is doubled, stars that are only one-quarter as bright may be photographed. The magnitude of the faintest star that may be recorded, the *limiting magnitude,* depends on and increases with both the aperture and the *f*-number. The dependence of limiting magnitude on these quantities is shown in Fig. 75. An *f*/5 camera will photograph stars about three magnitudes fainter than an *f*/1 camera. Cameras with *f*-numbers larger than *f*/5 will

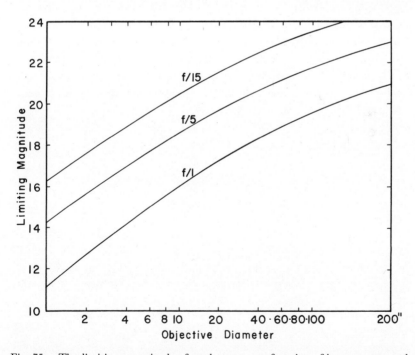

Fig. 75.   The limiting magnitude of a telescope as a function of its aperature and focal ratio.

not greatly increase the number of stars that may be photographed and the required exposure will become exorbitantly long.

We now see the other reason why low f-numbers are better for photographing nebulosities: the star images will not be so conspicuous. Also, if we decrease the size of the objective, keeping the same f-number, fewer and fewer stars will be recorded while the nebulosity will still be just as apparent, though of course on a smaller scale.

Thus either stars or nebulae may be selectively emphasized. One chooses a relatively long focal length and a large aperture to photograph the faintest stars, or a small f-number to minimize the presence of stars in a photograph of nebulae.

An important consideration, the permanence of the plate scale, enters into the choice of an instrument for astrometric work. Refractors are preferable for such work. Their optical elements are close together so that a rigid cell can maintain them in accurate alignment over many years. Though some of the newer catadioptric cameras can provide smaller star images for a given focal length and just as large a field, they cannot be depended upon to maintain accurate adjustment during the long periods required for parallax and proper-motion studies. This work is best performed with long-focal-length photographic refractors. As a result of these considerations many different types of instrument are found in observatories today and there is no one instrument that can perform more than a small part of the varied research of modern astronomy.

4

Construction
of Telescopes

Lens and Mirror Cells

Lenses and mirrors need to be permanently mounted in order to fix them in their proper position in the optical train. Small optical elements present no difficulty. Lenses are usually mounted in short tubes. Multiple-lens systems often have provisions that permit adjustment, such as centering and spacing, of the various elements. The adjustment of lens systems of large size is often critical and requires considerable effort and test tools. To reduce the need for taking the lens systems apart to clean the surfaces, the space between the elements should be kept so tight that dust cannot enter. However, rings and spacers holding the lenses in place must not exert a distorting pressure on the glass, which would lead to internal strain and stress, and from which, in turn, the quality of the images would suffer. Multiple-lens systems are often so sensitive to small changes

111

in the alignment that temperature changes, particularly when they do not affect the entire system uniformly, disturb the adjustment of the elements more than can be tolerated for optimum performance. Tolerances must allow sufficiently for the longitudinal thermal expansion of the cell, which causes changes in the separation of the lenses, and for the differences in radial thermal expansion of glass and metal parts.

Cells for large lenses usually consist of two sections. The inner cell holds the lenses; it is screwed into the outer cell which, in turn, screws to the tube of the telescope. If no means of adjustment of the inner cell is provided to coördinate the optical axis of the lenses with the geometric axis of the tube, the various parts must be made precisely enough to ensure coincidence of the two axes. Adjustable cells frequently employ an arrangement of push-pull screws in pairs, 120° apart, on a flange. Mounting of lenses, even large ones, does not present particularly difficult problems such as the designer faces when he deals with mounting large mirrors. The fact that lens cells, by necessity, support the lens only at its edge is one of the reasons that, beyond a certain size, lenses are not practical, because they would sag so considerably that their distortion would become intolerable. The largest telescope lens is the 40-inch objective of the Yerkes Observatory.

Mirrors not exceeding 25 inches in diameter do not need complicated cells. The glass disk is placed in a container strong enough to carry the weight of the glass without being distorted. Pads of cork or carpet are frequently put between the glass and the cell bottom.

The success of a telescope employing a large primary mirror depends upon how satisfactorily the designer has solved the problem of support for the mirror. The cell and the support system are among the most essential parts of a large reflector. They must maintain the shape of the reflecting surface with a high degree of accuracy, regardless of the orientation of the mirror and of temperature changes. Flexure of a mirror under its own weight is proportional to the fourth power of its diameter and inversely proportional to the square of its thickness. This indicates how rapidly flexure caused by gravity becomes a serious issue with increase of size. If the thickness is increased in proportion to the diameter (that is, if the ratio of diameter to thickness is kept constant), flexure due to gravity will increase only with the square of the diameter. This is a serious problem, especially since the total weight of the disk is great for big mirrors.

To reduce the weight of very large mirrors without impairing the rigidity, a ribbed structure for the back side of the disk has been adopted. The primary mirrors of the 200-inch Hale telescope and the 120-inch reflector of the Lick Observatory are samples of this solution (Fig. 76). But even with the weight reduced in this fashion, an elaborate support system is required to carry a large mirror in

Fig. 76.   The 120-inch Lick mirror, back side. (Courtesy Lick Observatory.)

such a way that the gravitational distortion of its surface does not affect the quality of the images. To achieve this purpose with a large mirror of correspondingly high resolving power considerable efforts have to be made. A simple three-point support system, adequate for small mirrors, is entirely insufficient for big ones.

Systems consisting of a large number of individual support units have been devised which balance exactly the force of gravity acting on the section of the mirror assigned to each of the units (Figs. 77 and 78). This is achieved by applying the proper components of force in every direction. If the orientation of the mirror with respect to the direction of gravity changes, the components of force applied must vary accordingly to compensate the gravitational force within 0.2 percent or better. A number of units, made of weights and compound levers, spaced properly for the weight distribution over the disk, are used to act against the rear surface of large mirrors. Similar devices are spaced equally around the periphery of the mirror; they act radially and support the mirror at its edge when it is tilted. The total weight of the mirror is divided between the two systems

Fig. 77.   Mirror cell and support system of 120-inch reflector, seen from behind. (Courtesy Lick Observatory.)

Fig. 78.   Mirror support unit of 120-inch reflector. (Courtesy Lick Observatory.)

and their individual units, depending upon the angle of tilt the telescope assumes. Support systems of this type are now widely used with large mirrors.

Recently, such a cell has been built for the mirror of the 61-inch reflector of the Harvard College Observatory. The weight of the mirror is balanced against the weight of the counterweights in each support unit by a pressure-sensitive hydraulic device consisting of two sylphon bellows of properly chosen diameter ratio and interconnected by a pipe. The bellows are filled with a hydraulic fluid to transfer the components of force.

The two systems of back and radial support do not define the position of the optic axis of the mirror. Radial definition of the mirror can be accomplished in various ways. If the mirror has a Cassegrain hole in its center, radial-position-defining bearings rigidly attached to the cell and acting against the periphery of the center hole can be used. If properly designed, they will obstruct the field of the Cassegrain focus very little. If the mirror has no center hole, a center-defining bearing can be fitted in an accurately ground hole in the back of the mirror. The center bearing allows the mirror to float axially without lateral motions.

The mirror—as well as the telescope tube—is subjected to changes of the ambient temperature. In climates with rapidly changing temperatures this occasionally constitutes a serious problem because the mass of the mirror is so large that it cannot adjust its temperature quickly and uniformly throughout its whole disk. A ribbed structure is advantageous because of the smaller mass involved, and because internal points are closer to the surface of the glass. Even with all precautions taken to ensure an equal flow of heat to all sections of the mirror, distortions due to temperature changes may occur occasionally. Unless provision is made for air circulation and proper insulation of sections of the mirror, the outer edge, for example, might assume a new temperature faster than the central zone, which would cause a distortion of the reflecting surface.

Temperature changes also vary the geometric distance from the mirror surface to the position assigned to the focal plane. The focal length of the mirror itself varies on account of changes of the temperature of the glass. An interesting innovation has been introduced in the construction of the 48-inch Schmidt telescope on Palomar Mountain to minimize the structural expansion effects. Three bars of Invar, a nickel-alloy steel with a very low coefficient of thermal expansion, are equally spaced around the tube. The lower ends of the bars terminate in spherical caps, adjustable for length, which rest against flat pads attached to the edge of the mirror by clamps. An auxiliary spring system attached to the axial supports maintains contact between the bar ends and the pads. The upper ends of the three bars rest against similar pads attached to the ring girder and blade assembly that supports the plateholder at the center of the tube. The ring girder, an extremely rigid member, is connected to the tube by flexible radial spokes. They are adjusted for equal tension. The Invar bars are firmly fixed to the tube near their midpoint. Above and below this point of anchorage, the bars are guided, free to move axially but restrained against any transverse motion. Since the length of the bars is practically independent of temperature, the distance from the mirror to the photographic plate is also practically constant.

Tube and Mounting

The tube of the telescope generally serves two purposes: to carry the optical components, plateholder and other auxiliary equipment properly spaced and aligned, and to prevent stray light from getting into the focal plane (Fig. 79).

Refractors, and frequently reflectors, too, particularly small and medium-sized ones, employ solid tubes. The tubes are occasionally double-walled to avoid flexure, especially when heavy auxiliary equipment such as large objective prisms are to be attached. For the construction of very large reflectors, the solid tube is often replaced by a framework. Special provisions have to be made to keep stray light from striking the photographic plate or the photocell at the focus.

Rigidity of the tube, regardless of the angle of tilt, is a very stringent requirement for photographic work. A slight displacement during the exposure of the plateholder with respect to the optical axis, as caused by varying flexure of the tube, produces trailed images of the stars on the photographic plate. Permissible tolerances are small, more so for instruments with long focal length. By a properly designed system of counterweights and levers, flexure can be reduced.

A very necessary accessory in locations with high humidity of the air is the dew cap. It is an extension of the tube beyond the objective. It greatly reduces condensation of water on the front surface of the lens, particularly if equipped with a heater which keeps the temperature of the air in front of the lens just slightly above the dew point.

Reflectors require a heavier construction than refractors of the same focal length, because of the larger weight involved. A framework of ring girders and truss members is lighter than conventional cylindrical tubes and, if properly designed, of high rigidity. Such a

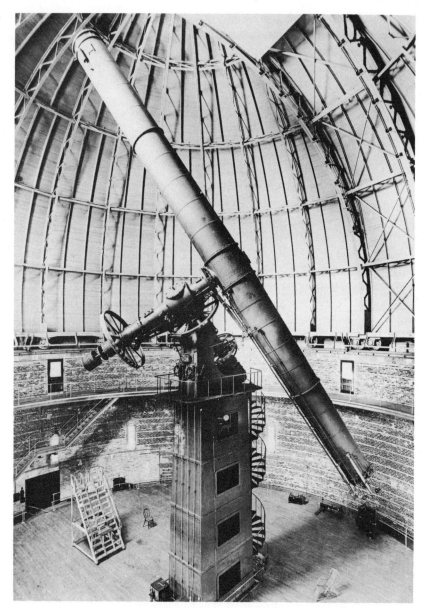

Fig. 79.   The 40-inch Yerkes refractor. (Courtesy Yerkes Observatory.)

framework consists frequently of a cubical middle section suspended from the declination axis. Bolted or welded to the center section are the frameworks that carry the mirror in its cell at one end of the telescope and the prime and Newtonian focus provisions at the other.

The prime-focus end also carries the observer's cage, if such an arrangement is provided, as on the 200-inch Hale telescope and the 120-inch reflector of the Lick Observatory (Fig. 80). The rigidity of such a framework can be made surprisingly high. The framework of the 200-inch including the cage and the mirrors weighs about 140 tons, but it does not sag more than $\frac{1}{16}$ inch.

A framework is preferable to a closed tube for a reflector because it favorably affects the image quality. If the temperature of the tube is not everywhere the same, turbulent motions of the air enclosed

Fig. 80.   Observer's cage at the front end of the tube of the 120-inch Lick reflector. (Courtesy Lick Observatory.)

might occur and reduce the image quality. Also, an open framework permits adjustment of the mirror temperature to the instantaneous ambient temperature faster than a mirror tightly enclosed in a tube can adjust.

Except for very special purposes, the telescope tube is supported by an equatorial mounting. This type of mounting permits the telescope to follow the daily motion of the star at which the instrument points by rotation around only one axis. The daily westerly motion of the stars is caused by the rotation of the earth around its axis. By rotating the telescope around an axis parallel to the axis of the earth, the daily rotation of the earth can be compensated in such a fashion that a star in the field of the telescope stays there. The ends of this axis, referred to as the polar axis, point to the celestial poles, where the earth's axis intersects the celestial sphere. Perpendicular to this axis is another one around which the telescope can be rotated, the declination axis. Thus the telescope has sufficient freedom to be pointed at any spot in the sky.

The way such a system of axes is executed in practice and supported on a foundation leads to a variety of modifications of the equatorial mounting. In 1618 Grienberger suggested a mounting which became quite common, particularly for visual refractors, after Fraunhofer employed it early in the nineteenth century for his astronomical telescopes. The polar axis of this so-called German mounting is supported close to its center of gravity (Fig. 81). To the upper end of the polar axis is attached the declination axis, while the lower end frequently carries a counterweight. The telescope tube is at-

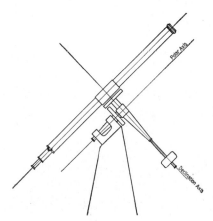

Fig. 81.   German equatorial mount.

tached at one end of the declination axis and a counterweight at the other. This system of axes is supported on a column or pier or—in the case of a small instrument—on a tripod. The column becomes an obstacle to the path of motion of the tube when the telescope points toward an object close to the meridian. If a star has a declination larger than the altitude of the celestial pole, one has to shift the telescope from the west side to the east side of the pier in order to follow the star when it passes the meridian. Similar difficulties arise for circumpolar stars at their lower culmination. Inconvenient as this is for visual work, it is not feasible at all for photographic exposures. In addition to the undesirable interruption of the exposure, the photographic plate would have to be rotated through 180° in order to continue. Greater freedom in the movements of the telescope is provided by a modified form of the German mounting in which the upper end of the column is made to overhang the lower part.

The English mounting is characterized by a long polar axis supported at both ends on piers (Fig. 82). The declination axis carries

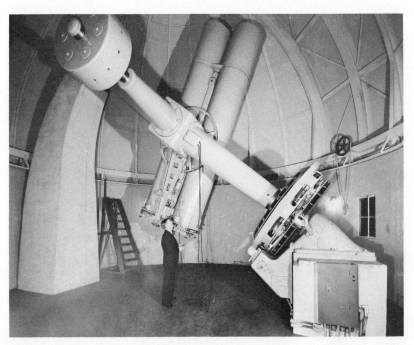

Fig. 82.   The 20-inch double astrograph, English equatorial mounting. (Courtesy Lick Observatory.)

the telescope at one end and a counterweight at the other. This type of mounting offers remarkably good rigidity and freedom from flexures; hence it is frequently used for long-focal-length instruments for photographic positional work. The distance between the two piers should be large enough to permit the telescope to swing between them. The English mounting is recommended for locations where the pole is at a low altitude.

Fig. 83.   Yoke mounting of the 100-inch reflector on Mount Wilson. (Courtesy Mount Wilson and Palomar Observatories.)

In a variation of the English mounting the tube is not overhung on a counterbalanced system of cross axes, but is suspended by trunnions in a rectangular frame, the so-called cradle or yoke (Fig. 83). The yoke replaces the polar axis and is supported at its north and south ends by bearings. The telescope is freely movable in the meridian and shifting from one side to the other of the polar axis is obviated. However, the telescope cannot be set on stars at or near the pole. For a very large instrument, such as the 100-inch reflector on Mount Wilson, the yoke is too short to accommodate the entire length of the tube; the tube cannot be fully swung through the cradle. This measure permits a reduction of the size and thus the costs of the dome, but with this mounting stars close to the north pole cannot be observed.

Some of the disadvantages of the types of mounts discussed so far are avoided by the fork mounting (Fig. 84). To the upper end of the polar axis a two-pronged fork is attached. The tube is suspended by the declination trunnions at the tips of the fork and swings between the prongs. This mounting is very suitable for reflectors, where the center of gravity of the tube lies close to its lower end because of the great weight of the mirror and the cell. Thus, only a comparatively short fork is required. The fork mounting offers the great advantage that the entire celestial sphere above the horizon is easily accessible with the telescope, so that no shifting of the tube about the pier is needed when the object passes the meridian.

A rather special mounting has been devised for the 200-inch Hale telescope (Figs. 85 and 86). It combines the features of the two variations of the English mount (yoke and two-pier support) and of the fork mount. The polar axis consists of a large fork with the prongs joined by a cylinder 46 feet in diameter. A segment of this cylinder is removed in order to allow the telescope to point at the north pole of the celestial sphere. The resulting horseshoe serves as the northern bearing of the polar axis. It rests on a pair of 28-inch square pads, fitted to the cylinder and recessed in the center. Oil is pumped into the recess under pressure sufficient to lift the telescope from the pads. The horseshoe then floats on an oil film a few thousandths of an inch thick. The bearing at the south end of the polar axis operates with such oil films, too. Although the moving parts of the telescope weigh more than 500 tons, the friction in these bearings is so small that a $\frac{1}{12}$-horsepower motor drives the telescope at the sidereal rate, and an even smaller motor would be sufficient.

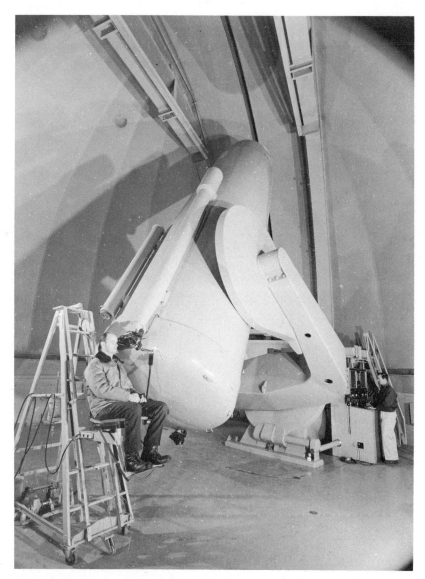

Fig. 84. Fork mounting of the 48-inch Schmidt telescope on Palomar Mountain. (Courtesy Mount Wilson and Palomar Observatories.)

Telescope Drives and Controls

An astronomical telescope needs to be equipped with driving mechanisms which offset the daily motion of the stars caused by the rotation of the earth. We distinguish three different kinds of motion that can be applied to the telescope: motion for pointing the tele-

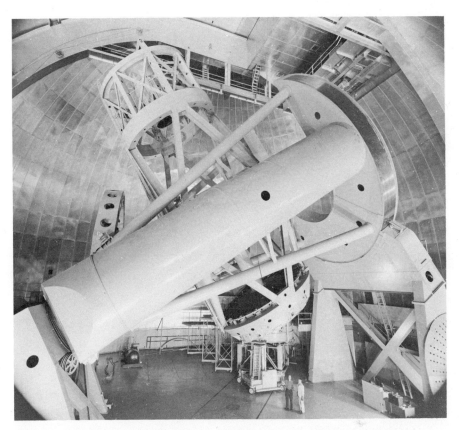

Fig. 85. The 200-inch Hale reflector. (Courtesy Mount Wilson and Palomar Observatories.)

scope in the direction of the object (setting), motion to make the telescope follow the object (driving or tracking), and motion to correct inaccuracies of the tracking motion by visual inspection or a photoelectric control device (guiding).

To move a large telescope rapidly into approximately the desired direction, motors are provided which can turn the telescope around its two axes at a conveniently fast rate. These "fast motions" move the instrument through 40° to 100° per minute of time, depending on the size of the telescope; small instruments are usually pointed by hand into the approximate direction. These motions are too rapid to permit setting the instrument on the star with sufficiently high accuracy. Two additional motors, one for each axis, controlled by the observer, move the telescope at a much lower speed from its approximate position until the object appears in the proper position

THE · TWO · HVNDRED · INCH ~
TELESCOPE · LOOKING · NORTHWEST

Fig. 86. Drawing by R. W. Porter of the 200-inch Hale reflector. (Courtesy Mount Wilson and Palomar Observatories.)

in the field. A typical speed for these "set motions" is 80′ of arc per minute of time, but considerably larger and smaller rates are occasionally provided.

The star would not remain in the field if the telescope were not driven around its polar axis with the same angular speed as that at which the star moves. This is accomplished by a motor whose output-shaft speed is properly reduced by means of gear trains (Fig. 87). The sidereal-rate motion is the heart of the telescope-drive mechanism. However, even if this tracking motion operated with the utmost uniformity of rate, corrections would still need to be applied. These corrections are required to compensate inaccuracies in the sidereal-rate drive mechanism itself, effects of atmospheric refraction on the apparent position of the stars, errors in the alignment of the axis of the mounting, varying flexures of the telescope tube, and other factors. These corrections to the tracking motion are applied by an observer who monitors the accuracy of the telescope tracking. It is done by observing, with the help of an eyepiece, a star image set on cross hairs in the focal plane of the telescope, on the jaws of a spectrograph slit, or in the field of an auxiliary telescope (guiding telescope) attached to the main telescope. The longer the focal length of the guiding telescope compared with that of the main instrument, the better is the accuracy achievable in guiding.

Fig. 87.   Section of the right-ascension drive of the 120-inch Lick reflector. (Courtesy Lick Observatory.)

The observer superposes correction motions upon the tracking motion by running the motors providing the guide motions. The rate of these motions usually lies between 0.1″ and 15″ of arc per second of time, depending on the size of the instrument and the particular purpose it is used for. Occasionally variable speed reducers are provided to make the rate of the guide motions better adaptable to the observer's needs.

Guiding of a telescope with a long focal length by means of an auxiliary telescope for taking photographic plates meant for accurate position determination of stars is sometimes inadequate because of differential flexure between the two telescope tubes. One way of eliminating the difficulty is to use a moving plateholder. The plateholder receiver is equipped with a guiding eyepiece. By means of two screws the plateholder and the eyepiece can be moved as a unit in the focal plane along two axes perpendicular to each other. To guide the telescope during the exposure the observer sets the eyepiece with its cross hairs on a star, usually at the edge of the field to be photographed, and moves the plateholder as required to accomplish proper guiding.

Numerous arrangements have been designed to generate the sideral-rate motion of the telescope. The oldest drive is the weight-driven clock as introduced by Fraunhofer and by Cooke and later modified by others. Its rate can be controlled by a mechanical governor. Winding of the clock is done manually or—with bigger telescopes—automatically by a motor. The rate of a governor-controlled clock cannot always be changed while the clock is running, an inconvenience at times.

Speed control of drive clocks is occasionally accomplished by the swinging pendulum of an ordinary clock. Willard P. Gerrish devised such a drive which makes the rate independent of voltage fluctuations (Fig. 88). The magnet A, attached to the pendulum, lifts the armature B and opens the switch C when the pendulum swings through its vertical position. As soon as C has broken the 10-volt control circuit, the armature H drops so far away from the electromagnet that it cannot be pulled back, even when the 10 volt circuit is closed again as soon as the pendulum swings past its vertical position. When H drops, it closes the power switch F and the driving motor starts. A cam J geared to the motor shaft returns the armature H to its original position, where it is held by the 10-volt magnet until the pendulum repeats the process on the return stroke. The

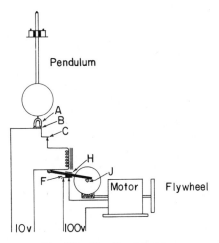

Fig. 88.   The Gerrish drive.

interval between the time the pendulum reaches the vertical position and the time the cam lifts thc armature constitutes the power
stroke of the drive. If the motor is running too fast the cam will shut
off the power soon after the pendulum releases the armature *H*. If
the motor is running too slowly, the cam will not open the switch
F so soon and the power stroke is longer. In a variation of the device
the magnet switch is replaced by a photoelectric cell and a beam
of light periodically interrupted by the swinging pendulum. The rate
of the clock can be changed by changing the load on the pendulum,
thus effectively shifting the center of gravity. A similar device has
been developed by M. Wolf in Heidelberg. Another speed-control
device stabilizes the speed of the motor with the aid of signals from
a clock by monitoring a mechanical governor, a method developed
by Zeiss.

Synchronous motors whose speed is determined by the frequency
of the supply voltage are becoming more and more common as drive
motors. However, although the line frequency may be constant
enough to maintain adequate tracking accuracy, the drive speed
cannot be varied easily to make the telescope follow an object, such
as a comet or a planet, that moves with respect to the stars, and cannot be adjusted to compensate for the effects of refraction.

Owing to atmospheric refraction the light rays coming from a star
are bent in such a fashion that the star appears a little higher in the
sky than it actually is. This elevation is larger as the star gets closer
to the horizon. When the telescope is pointed near the eastern or

western horizon, the drive must run more slowly than in the vicinity of the meridian where the change in refraction is small.

Modern drives frequently employ a synchronous motor supplied with alternating current generated by an electronic oscillator. The current put out by the oscillator is sufficiently amplified by a power amplifier to meet the power requirement of the motor. The frequency of the alternating current can be varied by changing the setting of a circuit component determining the frequency of the oscillation, such as a resistor or a capacitor. The rotor shaft of such a variable element can be turned by hand or by a motor. With a motor, the observer can conveniently control the drive speed from his remote post. This kind of drive is very flexible for adjusting the drive speed of the telescope to the speed of the object under observation and for changing the speed without interruption of the observation or the plate exposure, whenever the necessity for a change arises.

Occasionally the frequency of a vibrating wire is used to stabilize the drive speed. The frequency can be varied if desired by changing the tension of the wire.

The various motions of a telescope of large size are controlled by the observer with the help of a small portable block containing buttons or switches. This control block is fastened to the end of a flexible cable, permitting operation of the telescope from any observing position. For larger instruments, panels carrying the control elements are provided in addition (Fig. 89). The telescope can often be operated entirely from such a remote station by an assistant, while the observer mainly monitors the guiding.

The two axes of a telescope are furnished with divided circles by means of which it can be set on the object. Large telescopes are now often provided with electrical repeating devices by which the positions of the two axes can be read on dials at the control panel. Thus the assistant simply moves the telescope until the dials for right ascension and declination register the coördinates of the object.

A repeating system is established by the use of selsyns. Selsyns resemble 3-phase induction motors with two definite field poles. The windings of the field poles are connected to a single-phase alternating-current source of excitation. The shaft of one selsyn (transmitter) is geared to the telescope shaft; the shaft of the other (receiver) carries the dial. The secondary winding of the transmitter is connected to the one of the receiver. With the primary excitation circuit

Fig. 89.   Control desk of the 61-inch reflector of Harvard College Observatory.

an a-c voltage is impressed on the primary of transmitter and receiver. The receiver rotor turns until it has found a position similar to that of the transmitter. When the transmitter shaft is turned by the motion of the telescope the receiver rotor follows at the same speed and in the same direction. By combining several selsyns electrically and mechanically the reading accuracy of the dials can be increased in a fashion similar to a clock by providing hands for hours, minutes, and seconds (Fig. 90).

Selsyn systems as described transmit to the control panel the hour angle and the declination of the celestial position at which the telescope is pointed. For setting the telescope the observer must compute the hour angle, which is the difference between the sidereal time of the instant of observation and the right ascension of the object. Computation of the hour angle by the observer can be avoided if a selsyn differential system is provided.

A differential selsyn has three connections to the primary winding instead of two; it resembles a three-phase wound-rotor induction motor. If two selsyn transmitters are connected through a differen-

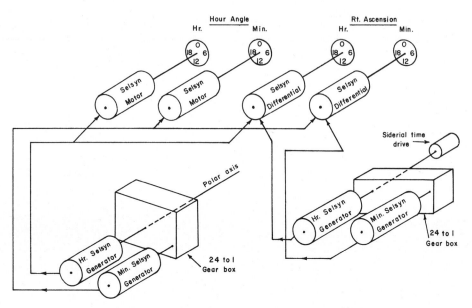

Fig. 90. Selsyn system for remote control of hour-angle and right-ascension set-ting of a telescope.

tial selsyn, and each of them is turned through any angle, the rotor shaft of the differential selsyn will turn until it indicates the differ-ence (or the sum) of the two angles. This can be utilized for repeating the right ascension of the telescope setting. Sidereal time and hour angle, or rather the corresponding position of the hour angle axis and the shaft of a sidereal clock, are fed into a differential selsyn with the help of selsyn transmitters. Then, the differential selsyn reads right ascension, and the observer simply moves the telescope until the dials of the right ascension and declination selsyns register the coördinates of the object.

Guiding of a telescope for a long exposure is a tedious task. Astronomers have made attempts to ease the strain on the observer by automatic devices for monitoring the motion of the telescope. Automatic guiders employing photocells are quite successfully used in solar instruments. For stellar work, satisfactory operation of such devices is difficult because of the low intensities of the light of the stars. A sufficiently bright guide star is frequently not in the field.

Photoelectric guide systems are finding more application for spec-trographic work with the largest telescopes. For manual guiding, the observer watches the stellar image on the slit of the spectrograph and corrects deviations from the desired position by actuating the guide motion. The image on the slit is seen through a small auxiliary telescope. In Babcock's photoelectric guider (Fig. 91) for the coudé spectrograph of the 100-inch reflector a short coaxial tube is mounted inside the eyepiece. Inside the tube, at the focal plane, an orange filter covering half the field is mounted. Its diametral edge accurately intersects the optical axis, around which the tube is made to rotate. The orange filter only partially obstructs the visible light but it is almost opaque for the light to which the photocell responds. If the image lies on the optical axis, the flow of light past the rotating filter edge will be constant. If the image is off the axis the intensity of the transmitted light will be cyclically varied. The light is directed to a photomultiplier, whose current output is amplified by an electronic amplifier. The amplifier output is conducted back to the rotating scanner where it is distributed to four separate circuits by a brush carried by the rotating tube. In this way, phase discrimination is established. The four contacts of the distributor are connected to four capacitors across which they develop voltages corresponding to the amount of starlight transmitted by the scanner in the four directions north, east, south, and west. These voltages are applied in opposing pairs to the grids of double-triode balance tubes with relays in the plate circuits. The contacts of the relays are uti-

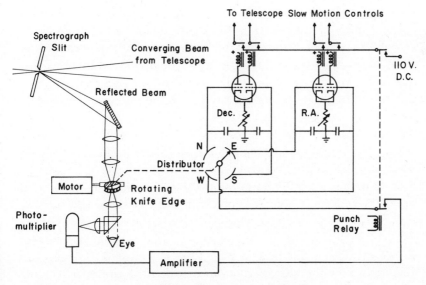

Fig. 91.   Photoelectric guider. (Babcock.)

lized to control the motors of the guide motions in right ascension and declination. The sensitivity of the device can be separately adjusted in the two coördinates.

Special Telescopes

For certain types of observational work, chiefly of astrometrical character, special support systems for telescopes are preferable to the equatorial mounting. If we disregard portable instruments and telescopes of a highly specialized kind, the most important telescopes for positional work at the present time are the meridian circle, the vertical circle, the zenith telescope, and their various modifications. These telescopes are sometimes referred to as fundamental instruments since they serve to provide the fundamental information on the coördinates of celestial objects. They are also indispensable for the determination of time and for the solution of various geodetic problems.

The meridian circle is a refracting telescope that can be rotated around a fixed horizontal east-west axis (Fig. 92). Thus the telescope can move only in the plane of the meridian. The horizontal axis rests on bearings supported by two piers.

In the focal plane of the eyepiece of the telescope a system of fine wires is arranged; frequently these wires are part of a micrometer attached to the telescope. The vertical plane through the center of the objective is called the collimation plane. The straight line through the center of the objective and perpendicular to the axis of rotation is called the collimation axis. Ideally, the central wire in the field of the eyepiece should be in the collimation plane, and the axis of rotation should be perfectly horizontal and exactly east and west. These conditions can never be met perfectly; the unavoidable deviations are referred to as collimation, level, and azimuth errors. If the instrument could be mechanically perfect and properly set up, a star passing the central wire would simultaneously pass the meridian. The observer notes the instant of passage by means of a sidereal clock; its reading is equivalent to the right ascension of the star. Readings of passages over other wires parallel to the central wire can be taken into account in order to improve the accuracy of the observation of the time at which the star crossed the meridian.

Attached to the axis of rotation are divided circles permitting readings of the altitude at which the star passed the meridian. With these readings the declination of the star can be deduced (Fig. 93).

The design and construction of a meridian circle requires extreme care, if coördinate determinations of high accuracy are intended. The prevention or reduction of flexure in the tube, of instabilities in the suspension of the instrument, of minute irregularities of the

Fig. 92.   A meridian circle; the instrument is in the process of being reversed in its bearings. (Courtesy Askania-Werke, Berlin.)

Fig. 93.   Declination circle of a meridian circle. (Courtesy Askania-Werke, Berlin.)

bearings, of slight motions of the piers supporting the bearings is imperative. Free circulation of air around the instrument, precautions against nonuniform heating by radiation from the walls of the building in which the instrument is housed, and careful selection of materials for the parts of the instrument, are safeguards against unduly large disturbances.

How stringent the stability and accuracy requirements are becomes obvious from some of the problems the divided circles offer. Their diameter is of the order of 20 to 40 inches. The division is usually provided on a silver or brass ring inserted in the disk close to its limb. Even a division made with the utmost skill and care is never completely free from errors. These errors can be empirically determined and then applied to the readings as corrections. Such a scale occasionally needs cleaning, which is usually done by polishing it with charcoal made of linden wood. Even a little force applied to the metal in polishing may be sufficient to affect the positions of the division marks engraved on the circle and to make a previous error determination useless! Recently proposed inserts of monel (a nickel-copper alloy with traces of iron, manganese, and some other elements) or palladium-platinum are preferable, since they are not subject to oxidation requiring cleaning and they are also more resistant to mechanical action such as rubbing or polishing.

With the help of microscopes, readings of the circle settings to a fraction of 0.1 second of arc are feasible. Photographic recording of the circle setting instead of reading by an assistant observer is becoming more and more common (Fig. 94).

Observations of the time of meridian passage can be improved by employing impersonal micrometers (Fig. 95). The observer does not note the instant of the star's crossing of a wire. Rather he sets a movable vertical wire on the star's image, bisecting it exactly. This wire is automatically driven through the field with the speed of the star. The observer applies minor corrections, if necessary. The driving

Fig. 94.   Photographic recording unit of a meridian circle. (Courtesy Askania-Werke, Berlin.)

Fig. 95.   Micrometer of a meridian circle. (Courtesy Askania-Werke, Berlin.)

mechanism actuates electrical contacts signaling the passage over the meridian to a time-recording device.

The vertical-circle telescope resembles the meridian circle. It also serves for measuring the altitude of a star above the horizon, but it is not restricted to observations in the plane of the meridian. Provision is made to rotate the instrument around a vertical axis into any azimuth desired. The extreme stability of the azimuth offered by the meridian circle is lost, but the instrument has advantages for certain kinds of positional work; for instance, the exact position of the zenith on the celestial sphere can be easily found.

An instrument quite similar to the vertical circle is the zenith telescope. The circles attached to the horizontal and vertical axes carry a crude division only sufficient for setting. The eyepiece is equipped with a micrometer, the screw of which moves a wire in the field. If the telescope points toward the meridian, this wire is perpendicular to the meridian and moves in its direction.

The zenith telescope serves for very accurate determinations of the geographic latitude and its variations caused by motions of the poles of the earth. The method was proposed by Talcott in 1834.

The telescope is pointed at a star passing the meridian close to the zenith, say south of it. The micrometer wire is accurately set on the star. After the star has crossed the meridian, the instrument is turned through 180° around its vertical axis. If the altitude at which the telescope is set has not been changed, it points now slightly north of the zenith. Another star passing through the field is observed in the same way as the preceding. The difference of the two micrometer readings is the difference of the zenith distances of the two stars. The latitude of the place of observation is the mean of the declination of the two stars, corrected for half the difference of their zenith distances. The method produces extremely accurate results because many instrumental and observational errors do not matter; for example, errors introduced by flexure of the telescope and effects of atmospheric refraction are largely eliminated.

Even higher accuracy can be obtained with the photographic modification of the zenith telescope. Absolute latitudes can be found accurately within a few meters, and latitude differences of locations even several thousand kilometers apart can be determined within 0.5 m. The photographic zenith telescope (Fig. 96) employed by the U. S. Naval Observatory consists of a tube rigidly fixed in a vertical position. Thus, only stars passing over the meridian close to the zenith can be photographed. The length of the telescope tube is approximately half of the focal length. At its lower end is a basin filled with mercury. The mercury surface represents the horizontal plane of the place of observation. It reflects the light from a star cast upon it by the objective lens back to the upper end of the tube. The light beam comes to its focus just underneath the lens, where a small photographic plate is located. Lens and plate are arranged so that both can be tilted as a unit through a small angle without altering the position of the image on the plate. If the plate and the lens are rotated through 180°, the distance on the plate between the images of the star, taken before and after reversal, corresponds to twice the zenith distance of the star. In practice, the two photographs of the star cannot be taken exactly when it passes the meridian. Therefore the plate is driven from west to east by an automatic mechanism and at the proper speed, and the time at which the plate is in certain positions is recorded. The geographic latitude of the instrument, the declination of the star, or the time of its passage across the meridian can be deduced, depending upon the aim of the measurements on the plate.

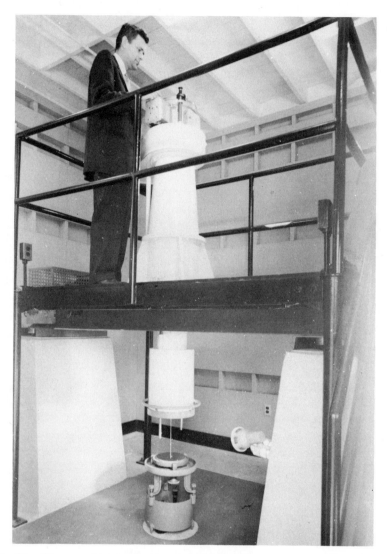

Fig. 96.   Photographic zenith tube of the U. S. Naval Observatory, Washington, D.C. (Official U. S. Navy photograph.)

Housing of Telescopes

Ordinarily, telescopes are kept in suitably designed buildings covered by a roof to protect them against the influence of weather and wind. For small-sized instruments no particular problems arise and fairly inexpensive shelters can be built. For large telescopes special constructions are indispensable.

Small instruments, such as little cameras of short focal length, may be sufficiently protected by a canvas cover that can be removed during observing, but strong winds and gusts may make even a small camera vibrate and thus prevent sensitive work. Also, the observer has no shelter against the wind, a serious disadvantage when he is expected to take long-exposure plates requiring accurate guiding.

A kind of housing often considered to be adequate for small-sized instruments is a construction employing a sliding roof. The box-shaped building is covered by a flat or gabled roof which can be slid off the building to allow operation of the telescope. Wind protection of the instrument is not very effective, in fact often quite unsatisfactory, even at low wind speeds.

A large telescope should never be housed in this way, especially when it is located on a mountain top or in an exposed place where strong winds are frequent. The surface area presented to the wind by a large instrument is so big that even light winds may make it vibrate sufficiently to affect the observations badly. A dome, usually of a hemispherical design, provides better protection of instrument and observer (Figs. 97–101). The costs of such a construction are

Fig. 97.   Lick Observatory of the University of California; dome of 120-inch reflector in foreground. (Courtesy Lick Observatory.)

Fig. 98.   Dome of the 24-inch reflector of Harvard College Observatory.

Fig. 99.   Dome of the 16-inch coronagraph of Sacramento Peak Observatory. (Courtesy Sacramento Peak Observatory.)

Fig. 100. Dome of the 20-inch double astrograph on Mount Hamilton. (Courtesy Lick Observatory.)

considerable and represent a sizable fraction of the expense of erecting a telescope of large size. A section of this hemispherical roof can be opened to give the telescope access to the sky. This slit is normally covered by one or two shutters which can be moved out of the way by hand operation or a motor drive. The slit ought

Fig. 101.   Dome of the 200-inch Hale reflector on Palomar Mountain. (Courtesy Mount Wilson and Palomar Observatories.)

to be wide enough to permit a light beam to enter the objective without vignetting. A slit somewhat wider than necessary to accommodate the beam is advantageous, but offers a larger entrance to the wind, too. Also, the slit should extend far enough beyond the zenith of the dome to allow unhindered observation of objects close to the zenith.

The dome must have provision for rotating in order to move the slit opening into the position required to observe the particular object. The dome turns on casters or trucks running on circular tracks. Small domes are manually rotated, medium- and large-sized ones are equipped with motors operated by the observer.

Automatic control of the dome motion coördinating it to the motion of the telescope is difficult to achieve. Only very few domes are equipped with a mechanism for this purpose. Since the relation between the motions of the telescope and the dome is rather complicated, no simple device for coördinating them is feasible. The world's two largest reflectors, the 200-inch and the 120-inch in California, synchronously control the motion of a so-called phantom telescope. This is a device simulating the motion of the actual telescope and the dome. Whenever the edges of the dome slit come too

Fig. 102.   Elevator in the slit of the dome of the 120-inch reflector on Mount Hamilton. (Courtesy Lick Observatory.)

close to the entrance beam of the telescope the phantom actuates the motion of the dome in the right direction and by a sufficient amount.

With a big dome the slit opening may be of a large area. To prevent wind from hitting the instrument, wind screens are occasionally provided. They can be moved up and down in the opening to cover sections not required to be open at the instant of observation.

During the day, the dome is exposed to the solar radiation, heat

Fig. 103.   Housing of a meridian circle. (Courtesy Hamburg Observatory.)

is absorbed, and the temperature of the interior is raised. During the evening and the night the warm air tends to leave the dome through the slit opening, while the temperature of the outside air drops. The turbulent motion of the warm air mixing with the cold has deleterious effects on seeing. In addition, rapid changes of temperature affect the telescope itself, alter distances between optical elements, cause changes of their focal lengths, and lead to poor performance; the problem is particularly serious if large masses of glass are involved, as in big reflecting mirrors. A heat-reflecting surface finish on the domes or insulation by double walls beneficially reduces the daily temperature changes of the dome interior. Double walls should permit free circulation of the air between surfaces and exchange air with the free atmosphere through louvers strategically located.

The dome structure, which occasionally carries elevators or platforms for reaching the observing stations (Fig. 102), must be designed to resist forces exerted on it by wind pressure and snowloads. The total weight of the dome of the 200-inch Hale reflector, having a diameter of 45 m, is approximately 1000 tons. Motions of such large masses, and also of considerably smaller ones, easily cause vibrations of the supporting walls which may be transferred to the foundations of the telescope piers and the telescope itself. This is, of course, detrimental to the observations, and precautions should be taken in the design to reduce the generation and transfer of such vibrations. The same ought to be done against vibrations caused on the floor of the dome, by machines, such as power generators in other parts of the building, and by nearby traffic.

Telescopes for special purposes, like meridian circles and zenith telescopes, do not call for such elaborate and costly constructions. They can be pointed only at limited sections of the sky, the meridian or the neighborhood of the zenith. A stationary and properly located slit in the roof of a suitable building serves the purpose (Fig. 103). Reduction of daily temperature variations inside the building by careful design and wise choice of materials is essential for obtaining results of high precision.

5

Photometry

Magnitudes and Colors

The oldest document we know of that contains information on the brightness of the stars is a book by Claudius Ptolemy (A.D. 137). It is known as the *Almagest,* a distortion of the Arabian translation of its title, *Al Majusti.* In a catalog of 1022 stars Ptolemy introduced the magnitudes 1 to 6 in order to describe their brightness. There was an even older star catalog by Hipparchus (129 B.C.) with data on the brightness of the stars; the work is lost, but it seems that Hipparchus divided the stars into only three classes with respect to their brightness. An important revision and extension of Ptolemy's work on the magnitudes of stars was accomplished by the Persian astronomer Al Sufi in the 10th century. He reëstimated and improved Ptolemy's magnitudes, increasing the total number of stars observed to 1151.

In the following centuries various astronomers pursued the task of determining stellar magnitudes. Tycho Brahe and Johann Bayer, the latter the author of a famous catalog, *Uranometria* (1603), were concerned with the problem, as were many others. Galileo Galilei extended the magnitude scale to faint objects which he was able to see with his telescopes. About 1830 remarkable progress was achieved by Sir John Herschel, who noticed that the arithmetic progression of the magnitudes is associated with a geometric progression of the apparent brightness of the stars. The ratio of the brightnesses of two stars of one magnitude difference in brightness turned out to be fairly constant through the whole scale and close to 2.5. The constancy of this ratio does not come as a surprise. It is an example of a general psycho-physical relation discovered by Weber and more precisely formulated by Fechner in 1859. Fechner's law of sensation states that differences in intensities which correspond to the same fractional part of the whole are perceptible to the senses equally well, no matter whether the original intensity is large or small. Pogson, in 1856, proposed the adoption of an intensity ratio of 2.5119 [$= 100^{1/5}$] for a magnitude difference of 1.00, and he suggested further that the zero point of the scale should be adjusted so as to secure as good an agreement as possible with the data in the earlier catalogs for the sixth magnitude. This scale with Pogson's ratio as constant is now extended to faint stars all the way to the twenty-third magnitude, which have been photographed or observed photoelectrically with the most powerful telescopes.

The mathematical relation between brightness and magnitude is simple. If we remember that a first-magnitude star is 2.5119 times as bright as one of the second magnitude which, in turn, is 2.5119 times as bright as one of the third magnitude, we can write in a more general fashion for the brightness ratio l_1/l_2 of two stars with the magnitudes m_1 and m_2,

$$l_1/l_2 = 2.5119^{m_2 - m_1},$$

or, if we rewrite the equation in logarithmic form,

or
$$\log (l_1/l_2) = 0.4 (m_2 - m_1)$$
$$m_2 - m_1 = 2.5 \log (l_1/l_2).$$

The following table shows the relation between difference in magnitude and brightness ratio.

Magnitude difference	Brightness ratio
1	2.512
2	6.310
3	15.849
4	39.811
5	100.000

We conclude that it requires the total light of 100 stars of the sixth magnitude, or almost 40 of the fifth magnitude, to equal the brightness of a star of the first magnitude.

In our discussion it has been implicitly assumed that the magnitudes were derived by visual estimates of the brightness of the stars or perhaps by a photometric device in which the human eye serves as a light detector (visual photometer). Now, it is well known that the eye is not equally sensitive to light of every color. Its response curve has a well-pronounced maximum for green light, at about 5500 A. It is not sensitive at all to light with a shorter wavelength than about 4000 A or longer than 7000 A.

The colors of the stars vary quite widely, ranging from blue through white and yellow to a deep red. Furthermore, we can easily see that the magnitude is also determined by the color-response curve of the light detector, which does not necessarily have to be the human eye; it might be a photographic plate or a photoelectric cell.

An example will serve to illustrate this point. On a photograph (Fig. 104) of a stellar field made with an ordinary blue-sensitive emulsion, we might find two stars that have produced exactly equal images. We conclude that the two stars are equally bright as seen by a blue-sensitive emulsion. Suppose now that one is a blue star and the other a red one. On a photograph taken with a red-sensitive emulsion the red star will produce a strong image while the image of the blue star is much fainter. The red-sensitive emulsion sees the two stars as of different magnitude whereas the blue-sensitive one does not.

The fact that light detectors such as the eye or the photographic plate lead to different magnitudes of the stars depending on their color is widely utilized and has led to the introduction of a quantity called the color index.

The color index of a star is simply the difference of its magnitude as measured by two light detectors of different color responsiveness.

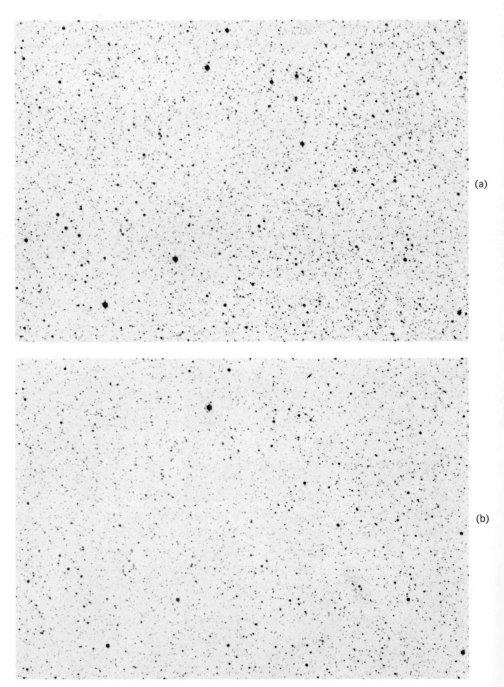

Fig. 104.   Negative prints of the same stellar field, taken with (a) a blue-sensitive, (b) a red-sensitive emulsion.

Frequently, the selection of the color range in the spectrum in which the measurements are taken is made with the help of color filters. They transmit light only in a limited range of color or wavelength. These filters are usually made of glass or gelatin. Recently, interference filters with narrow-band transmission have become important for special applications in stellar photometry.

The color index of a star is of considerable astrophysical interest. The surface temperature of the star determines the color of its integrated light. Consequently, the color index can be interpreted as a measure of temperature (Fig. 105). Color indices measured in strategically selected wavelength regions of the spectrum can yield a variety of important information on the physical conditions in stellar atmospheres and also in interstellar space, between the stars and the earth. Diffuse matter, thinly distributed in space, absorbs, scatters, and reddens the stellar light passing through it in the same way that tiny particles suspended in the terrestrial atmosphere redden the light of the sun and the moon, particularly when these bodies are close to the horizon. The amount of interstellar reddening can be found by a careful analysis of the color indices of the stars.

By using different combinations of filters and light detectors an unlimited number of magnitude and color systems can be set up.

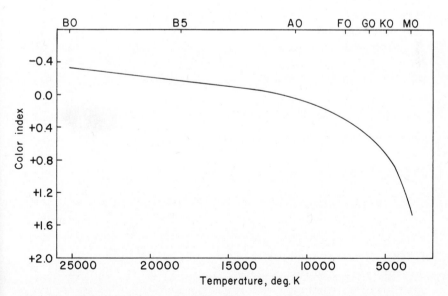

Fig. 105.   Relation between color index (International System), temperature, and spectral type for main-sequence stars.

Two such systems have been preferred for general use. These two systems and the corresponding color indices have been adopted by international agreement. They are the International Photovisual and the International Photographic Systems. The photographic system is roughly the one produced by an ordinary photographic plate most sensitive from the blue down to the ultraviolet limit of atmospheric transmission. Orthochromatic plates combined with yellow filters display a color sensitivity very similar to that of the human eye. Magnitudes determined with such a combination are correspondingly called "photovisual" magnitudes.

In recent years some special wavelength regions of the spectrum have been selected for magnitude measurements in order to obtain color indices permitting a clear interpretation in astrophysical terms. Also, with the introduction of highly accurate photoelectric methods for measuring magnitudes of stars, special photometric systems have been established. Two such systems are the P,V-system (the letters P and V are reminders of the older terms, photographic and visual, to which the photoelectric magnitudes of this system are fairly close) and the U,B,V-system of three different magnitudes in the regions ultraviolet (U), blue (B) and yellow (V for visual).

The zero-point of most of these systems is so adjusted by convention that the differences between them (the color index) becomes zero for certain white stars of spectral class A0.

Magnitude and color systems are in practice defined by groups and sequences of standard stars for which these data have been determined with the utmost accuracy. Such standard stars can be found in the so-called Polar Sequence—a group of stars around the celestial north pole—and also in some stellar clusters and various selected areas of the sky. Magnitudes are usually determined by comparison with these standard stars.

Visual Photometry

Visual photometry is by far the oldest branch of observational astrophysics. Data on the magnitudes of stars in the early catalogs were not measured with the help of instruments devised for the purpose but were obtained by visual inspection and estimation on a memory scale, or by comparing the stars with standard comparison stars of known magnitude. It was not until the 18th century that observers began to use measuring instruments, called photometers.

In 1725 Bouguer compared the sun and moon with a candle. Celsius and Tulenius in 1740 used a primitive photometer on 64 bright stars. Progress in the construction of photometers was rather slow for almost 100 years, but from 1830 on numerous designs became available. They were mainly based on two different principles: extinction of the star to be measured, or adjustment of its brightness to that of another star or to an artificial light source.

In a photometer of the extinction type a suitable component would permit reducing the brightness of the stellar image steadily until it would become invisible. The light-reducing device carries a scale. The reading of the scale taken at a position where the star image disappears serves as a measure of brightness.

For reducing the light, a calibrated wedge inserted in the beam is frequently used. Its density varies continuously or in steps from one end to the other so that the fraction of the stellar intensity that it transmits depends on the section of the wedge where the beam passes. Wedges covering an interval up to nine magnitudes can be made. Usually they consist of pieces of dark, neutrally absorbing glass ground to the desired shape. Cemented to the dark wedge is a wedge-shaped piece of clear glass arranged in such a manner that the two form a plane-parallel plate. The absorption of this plate increases from one end to the other. A photometric "wedge" can also be made of a piece of glass on which a coat of metal, say platinum, has been evaporated in such a fashion that the transmission varies at the desired rate from one end to the other. The absorption of such wedges, whether made of dark glass or a metal coat on glass, has to be determined by calibration. It is essential that it shall not depend upon the color of the light, a requirement that is hard to satisfy.

Another group of visual photometers compares the intensity of a stellar image with the intensity of another one or of an "artificial star," usually the image of a small opening illuminated by a standard light source. The intensity of one or the other is reduced with the help of a wedge or a polarizing element until the eye decides that the two images are equally bright. The amount by which the intensity of one image has to be reduced in order to match the intensity of the other is a measure of the difference in brightness of the two stars or of one star and the artificial source.

There is a large variety of modifications of the basic principle of the so-called comparison photometers. Important instruments have been developed by Zöllner, Pickering, Müller, Graff, and others. Their basic features are rather similar. We shall briefly outline the general scheme of a Zöllner comparison photometer equipped with a polarizing element (Fig. 106). The artificial star is produced by a diaphragm illuminated by a light B. The negative lens m and the positive lens f form two images of the diaphragm in the focal plane of the telescope after the beam has been reflected at the front and the back surface of the transparent glass plate ee'. In addition to the two images of the artificial star, the image of the real star appears in the eyepiece E of the telescope, produced by the objective O.

The beam coming from the light source has to pass through two Nicol prisms i and h, the first of which can be rotated while the second remains fixed. The intensity of the artificial stars depends on the relative position of the two Nicols, acting as a polarizer and an analyzer. After the intensities of the images of the real and the artificial star have been made equal, the angle through which the Nicol i has been turned is read and transformed into a magnitude difference. Thus the magnitude differences between various stars can be derived.

If the two Nicols and the circle carrying the angular scale are properly adjusted, the magnitude difference corresponding to two readings can easily be found from a table computed with the help of trigonometric functions. The light of the artificial and of the natural star may be fairly different in color, and the observer may then find it rather difficult to decide when equality of intensity has been reached. An auxiliary element, a "blue wedge" with variable color density or a simple color filter, can be inserted in the beam to match the colors as closely as possible.

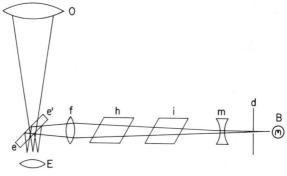

Fig. 106.   Principle of Zöllner's visual photometer.

The accurate measurement of large magnitude differences with a polarization photometer is difficult and requires considerable experience. Wedge photometers are easier to handle, and many observers have preferred them because of this advantage. Their construction does not differ much from the photometers employing polarization devices. The set of Nicol prisms is simply replaced by a photometric wedge which can be moved in the beam until the eye finds that the intensity difference between the images has been removed. Magnitude differences are obtained from a calibration curve or table, with the position of the wedge read on a scale as argument. Photometric work employing visual photometers is rarely done today. Photographic techniques and modern photoelectric methods have made visual photometers entirely obsolete. Some of the photoelectric means for measuring stellar brightness, particularly since the photomultiplier tube has become available, are so simple and reliable that now even amateurs use them successfully.

Photographic Photometry

An exposed photographic plate constitutes a lasting record of the brightness of many stars at a given instant and is always available later to check errors in the measurements. The brightnesses of thousands of stars can be determined on a single plate exposed for perhaps an hour or less. The photographic plate permits the most economical use of valuable telescope time. Another of its advantageous features is its ability to accumulate the light that falls on it during the exposure. Thus, much fainter objects can be reached for photometric measurements than can be seen and measured visually with the help of the same telescope.

The density of the photographic image depends on both the intensity of the light focused on the plate and the exposure time. Stellar images would be precisely luminous points in the focal plane if they did not assume finite size because of a number of circumstances, such as the effects of atmospheric seeing, diffraction by the finite aperture of the telescope, and imperfection of the optics. If the image is produced on a photographic plate, various processes bearing on image formation in the emulsion contribute to the finite size of the image (Fig. 107). The amount by which these individual processes enlarge the diameter of the final image on the photographic plate depends on the brightness of the star.

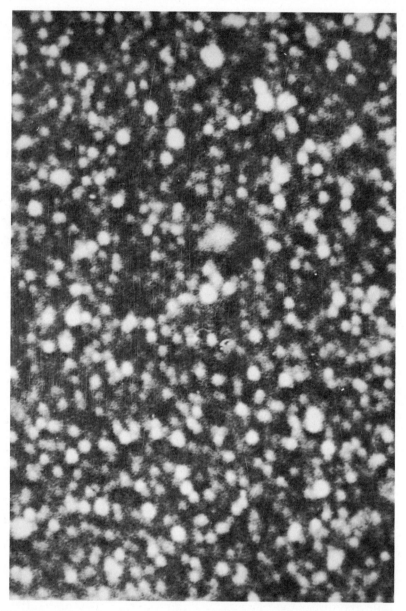

Fig. 107.   Enlarged photographic stellar images in a crowded field, taken on an Eastman 103a-0 plate with the Armagh-Dunsink-Harvard telescope of the Boyden Observatory.

The first problem to be solved in photographic photometry involves the finding of a relation between the density, or size, or "strength" of the stellar images on the plate and the intensity or the magnitude of the stars that produce the images. This is accomplished by photometric methods to some extent analogous to the ones devised for visual measurements. Instead of looking at the stars in the sky, the photometer looks at their images on a photographic plate, and the human eye or a photocell serves as the discriminating detector, much as it does for direct visual photometric work.

The use of the photographic plate causes complications that are bound to be sources of additional errors in the already delicate procedure of measuring the intensity of faint light sources under rather unfavorable conditions. There is the nonlinearity of response of the plate, strong variations of sensitivity with color, and a large number of factors hard to control and consequently quite variable in their effects.

The errors imposed on the measurements as a result of the photographic process can be kept to a tolerable minimum only by exercising the utmost care, precision, and skill in handling the exposure, the development of the plate, and the evaluation of information recorded in the emulsion. Considerable attention has to be paid to various other factors, such as seeing and transmission of the atmosphere during the exposure. Only under these conditions can we secure the most essential prerequisites to obtain reliable data. With all possible precautions photometric results accurate within 2 or 3 percent are feasible, and this is certainly the best we can hope for; often such precision will not be reached, but even then the results might be acceptable for many purposes. The advantages of the photographic process outweigh the disadvantages.

Information on the brightness of a star can be obtained by visual inspection of the images on the photographic plate. We might want to compare the density, the size, or the general appearance of the images. If we know the magnitude of some standard stars in the field we can find the magnitude of the stars to be investigated by a process of interpolating their images into the sequence of images of the standard stars. This is not too difficult and an observer with some experience can visually estimate magnitudes in this fashion with an accuracy of about 10 percent or a little better.

The images of the standard stars should appear in the field of the microscope used for the inspection simultaneously with the image of the star under consideration. This cannot always be achieved because the linear distance between the images on the plate may be too large. A step scale—sometimes called a "fly-spanker"—is conveniently applied in this case. It is a sequence of stellar images obtained by making a series of successive exposures with durations varied by a constant ratio. The plate is moved slightly between the exposures. This scale appears in the field of the microscope together with stars to be "measured." The images of the stars under investigation are then estimated in this scale. The images of the fly-spanker can be calibrated against stars of known magnitude and the photometric information on stars in the field of the plate can be derived by interpolation.

If the field does not include standard stars of known magnitude, another region of the sky which contains an adequate number of them can be exposed on the same photographic plate after the photograph of the selected field has been taken. Such a tie-in requires absolutely identical atmospheric conditions of the two regions and identical exposure times. It is advisable, if possible, to expose the two regions when they have equal altitude above the horizon. Then one can expect that seeing is of a similar quality for the two exposures. Also, the reduction in brightness of the stellar images due to atmospheric absorption and scattering can be expected to be similar for the two exposures.

The exposure of the two fields on the same plate assures completely identical treatment in processing, but, even if seeing and transmission by the atmosphere are the same for the two exposures, the procedure is not perfectly satisfactory. The night sky is not entirely black. The general background sky between the stars is somewhat luminous owing to scattered light from stars or the moon, light of faint stars not distinguishable as such on the photographs, airglow (emission of light by the higher regions of the atmosphere), the light from the aurora borealis or australis, and light from terrestrial sources.

Preëxposure and postexposure affect the results. The images of the second exposure are built up on an emulsion that has been preexposed to the light of the general background, and the background light during the second exposure is superimposed on the images of the first exposure. Photometric errors introduced by these superimpositions can be prevented if one makes the exposures of the two regions on two separate photographic plates from the same box. The two plates get the same treatment in processing and, being taken from the same box, have emulsions of exactly the same quality.

It has already been remarked that the fundamental problem of photographic photometry consists in finding the plate response as a function of stellar magnitude. The process of the formation of a stellar image on a photographic emulsion is a very complex one. Various features of the images can be used as brightness indicators, such as the density of the central region, the diameter, or something even less well defined and occasionally called image strength. This quality is fairly close to what the eye might consider as an indicator for the brightness of the star that has produced the image.

Measuring devices used in photographic photometry respond to different properties of the image; but to which features of the image they actually respond is sometimes ill defined. Some, called densitometers, simply measure density, the human eye serving as a detector. Densitometers are comparison devices, by which the observer decides by visual inspection when the features of the two images under comparison reach equality.

The devices employing the eye fall into two basic groups, as indicated in Fig. 108. The light from a source L is brought to the eye by two different paths with the help of the mirrors S_1 and S_2. On the path by way of S_2 it passes through the photographic plate P; on the other path, by way of S_1, it produces the comparison beam. In the glass cube V (a Lummer-Brodhun cube) two fields are separately illuminated by the two beams, and the observer can look at them simultaneously. Sometimes a "chopping" mechanism replaces the Lummer-Brodhun cube and presents the two fields alternately to the eye.

The illumination of the two areas in the cube can be equalized

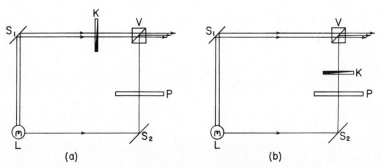

Fig. 108.   Two basic measuring schemes for photographic photometry.

by means of a wedge K inserted in the beam. The wedge can be moved in a measurable manner. The two arrangements shown schematically in Fig. 109 differ somewhat: in (a) the equality of the illumination of the two fields is reached at different light levels, depending upon the density of the plate P, whereas in (b) the reading is always obtained for the same light level, and errors that might depend on the illumination cannot occur. A comparison photometer of type (a) was suggested by J. Hartmann in 1899. It has been widely used and has served as a prototype for many modified devices.

Instead of using the eye as a light detector, a photocell can be placed behind the chopping mechanism. The photocell acts as a null indicator by utilizing the two beams, which are alternately imaged on the cell. If their intensities are equal, the photocell output is steady with no flicker at the chopping frequency. If the beams are not of equal intensity the photocell output is an alternating current of the chopping frequency. This ac-signal is amplified and supplied to a balancing servo motor that moves the wedge until the intensities of the two beams are matched. Then a reading of the position of the wedge is taken.

Photocells put out signals that are determined by the illumination, a property permitting construction of absolute measuring devices (Fig. 109). The photocell E produces a current proportional to the intensity of the light beam coming from the source L and passing through the photographic plate P. The density of the plate at the spot where the beam passes determines the photocurrent.

All these instruments employ a diaphragm which defines the width of the beam passing through the plate. A given diaphragm might be large enough for seeing the full image of a bright star. The image of a faint star would then cover only a small portion of the area of the diaphragm, and most of the beam would pass unaffected by the stellar image through the clear plate surrounding it. This

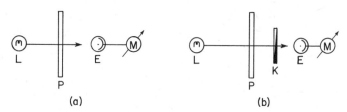

Fig. 109.   Two schemes for photoelectric photometry of photographic plates: (a) for absolute measurements; (b) compensation type.

would unfavorably influence the measuring accuracy. A diaphragm large enough for the strong images might not be permissible in regions crowded with stars, as in the Milky Way. A small diaphragm would be advisable for faint images or crowded fields, but it would not be wise to use it for strong images because only their inner cores would register. The accuracy of the measurements would suffer again, this time at the bright end; in addition, it is hard to set the diaphragm properly on the stellar image with its fringes cut off.

In 1934, Siedentopf suggested the use of a variable iris diaphragm to accommodate the width of the beam to the diameter of the image of each star. The aperture of the iris is adjusted in measuring so that the same amount of light passes through regardless of how large the aperture is. A photocell indicates balance. The quantity characterizing the brightness of the star is a reading of the setting of the diaphragm.

Two very desirable requirements are fulfilled in this device: (1) the larger star images do not spill over the edge of the image of the iris at the plate nor do the images of the faint stars cover only a small fraction of the beam area; (2) the photocell always works at the same level of illumination. It is wise to design the instrument in such a fashion that changes in brightness of the source are canceled out. This can be established by having a second beam from the same source for monitoring.

The basic system of a simple iris photometer as proposed by J. Cuffey is outlined in Fig. 110. It is largely based on a construction by H. Haffner, but somewhat simplified by the incorporation of modern electronic components. Two beams leave the lamp. The measuring beam passes a collimating lens, the iris diaphragm, and an objective lens which forms an image of the iris on the emulsion of the photographic plate. After passing the plate, another objective lens projects an image of the illuminated spot of the plate on a viewing screen for visual inspection, identification, and setting of the star image into the center of the beam. The mirror acts as a beam splitter. Part of the light continues through the mirror to a lens which throws it on the light-sensitive cathode of a photomultiplier. The monitor beam is also brought, with the help of some lenses and mirrors, over another path to the cathode. A chopper in front of the photocathode alternately permits only one beam to reach the cell. Both beams thus produce square pulses of different heights on the screen of an oscilloscope because the illumination of the cathode of

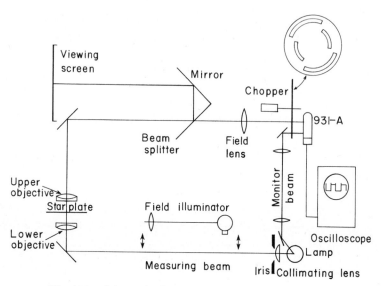

Fig. 110. Schematic diagram of Cuffey's iris photometer.

the photocell by the two beams is not identical. The illumination can be matched by varying the aperture of the iris; equality is indicated by a leveling out of the square pulse pattern on the oscilloscope screen into a straight line. When the balance stage has been reached a reading of the iris aperture is taken. In this fashion, magnitude ranges can be covered that are three or four times as large as those obtained with instruments employing constant-size diaphragms.

Contour Photometers

Occasionally, there is the need to study the brightness distribution of an extended celestial source such as a planetary nebula or some other nebulosity displayed on a photographic plate. To describe the brightness variations over the area of the object, contours of equal density corresponding to lines of equal brightness are frequently employed. Such lines can be set up by measuring the density of the plate at a sufficiently large number of places, a tedious and time-consuming undertaking if extended areas with complex structure are under consideration.

Several photoelectric scanning devices for reproducing isophotal contours from photographs have been constructed. A simple instrument proposed by H. W. Babcock may serve to illustrate the basic principles of contour photometers (Fig. 111). The essential components are two oscilloscopes, one of which, the scanner, has a short-persistence cathode-ray tube, while the screen of the other, the contour reproducer, should have long persistence. The two oscilloscopes are electrically connected in such a fashion that the horizontal amplifier of one drives the horizontal deflecting plates of both instruments, while the horizontal amplifier of the other drives the vertical deflecting plates of both. With appropriate frequencies for the internal sweep generators of the two oscilloscopes the luminous spots on the screens will synchronously scan similar rectangular areas. The scanning lines will be interlaced by making the sweep frequencies incommensurate and the whole screen will have been uniformly covered after many frames of scanning.

A lens at some distance in front of the scanning tube focuses an image of the screen of the tube on the photographic negative to be scanned. A field lens close behind the negative focuses an image of the first lens on the photocathode of a multiplier. The multiplier output is fed into an amplifier and a pulse-forming circuit which produces a positive voltage pulse whenever the image of the scanning spot crosses a portion of the negative that has the selected density level, in either direction. These pulses are then applied to the input of the intensifying amplifier of the reproducing oscilloscope, where they intensify the otherwise dim spot on the screen of the long-persistence cathode-ray tube, thus building up contours representing isophotes on the photographic negative. Selection of the density

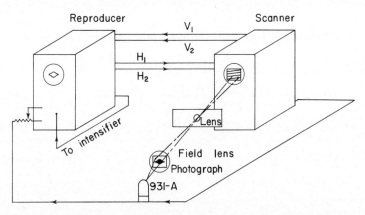

Fig. 111. Babcock's contour photometer. (Reproduced from the *Publications of the Astronomical Society of the Pacific.*)

level on the negative to be reproduced as a contour on the second oscilloscope is made by adjusting the intensity of the first scanning spot and by the use of the centering control of the signal amplifier.

The contours reproduced on the screen can be photographed by an ancillary camera for a permanent record.

Photoelectric Photometry

It is a well-established fact that the electrical properties of matter frequently vary with the illumination to which they are subjected. The dependence of these properties on light is a phenomenon that can be utilized to determine the intensity of light cast on a surface made of a responsive material. The nature of these photoelectric effects differs with the kind of material. Two photoelectric phenomena have found application in astronomical photometry: the photoemissive effect and the photoconductive effect.

Photoemissive materials emit electrons as the result of the impact of radiation while photoconductive materials change their electrical conductivity under the action of radiation. An early sample of a photoemissive cell is the potassium cell, in which potassium has been deposited on a layer of silver. A considerably more sensitive cell employs a cesium-antimony surface. Lead sulfide and lead telluride are substances that display the photoconductive effect.

Stellar light intensities are small and so are the electric currents produced by photocells under the influence of the faint radiation of stars. String electrometers for measuring currents were used for a long time; later the Lindemann electrometer replaced them. Usually, electrostatic methods served to measure the photocurrent. In the constant-voltage method the time required for the current to charge the electrometer to a fixed potential is measured. In the constant-time method, the potential attained in a fixed time is taken as a measure of the current. A variant of the last procedure is the steady-deflection method, in which the electrometer is charged by the photocurrent until a constant reading is attained which is proportional to the current. Skilled observers, like Guthnick in Germany and Stebbins in the United States, have developed photometric methods employing photocells. They have produced admirable results in spite of the tremendous difficulties involved in the practical application of a basically simple technique.

A major source of trouble in this kind of work stems from the atmospheric humidity, which frequently causes leakage of charge or breakdown of the insulation. The necessity of using resistors up to 10^6 megohms to obtain a measurable voltage drop with the extremely small photocurrent presents a serious insulation problem under working conditions of the kind to which a stellar photometer in an unheated room is exposed. Progress in increasing the sensitivity of the cells was slow. In 1932, A. E. Whitford succeeded in constructing an electronic amplifier for the small currents put out by photocells. A vacuum tube with a very low grid current, FP-54, of the General Electric Company, operating at low plate potential made possible significant amplification of these very low currents. A gain of about three magnitudes toward fainter stars was achieved by Whitford's amplifier.

In 1936 P. Goerlich proposed photocells with cesium-antimony cathodes which provide a remarkable increase in cell sensitivity, but the biggest improvement in photoelectric photometry was the introduction of the photomultiplier at the end of the Second World War. A device of this kind, the 1P21 of the Radio Corporation of America, has become the most widely used detector for stellar photometry. The primary photocurent is amplified within a sealed glass container by secondary emission. A multiplication factor of 10^6 is entirely feasible. The current put out by such a device is much more conveniently measurable than the one produced by an ordinary photocell.

Multiplication of the primary photocurrent of photoemissive cells had already been accomplished in a different way in gas-filled photocells. The gas introduced, argon for example, is one that does not react chemically with the photoelectric material. The electrons emitted from the cathode surface are accelerated by an electric field between the electrodes of the cell to a velocity high enough to enable them to eject other electrons from the atoms of the gas by collision. These secondary electrons and the primary ones may gain enough kinetic energy in the field to produce still more electrons by collision on their way to the positive electrode, thus increasing the current in the cell quite considerably (Fig. 112).

Multiplication of the primary photoelectrons in photomultipliers, however, is not done by collisions with atoms of a gas, but by focusing the electron beam released from the photosensitive cathode on a suitable metallic surface (Fig. 113). The number of secondary electrons produced here depends on the speed of the striking pri-

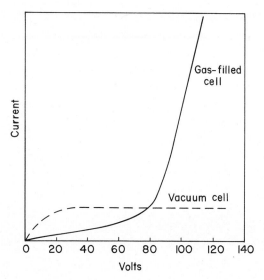

Fig. 112. Characteristic curve of a gas-filled and vacuum photocell, current output as a function of voltage across electrodes.

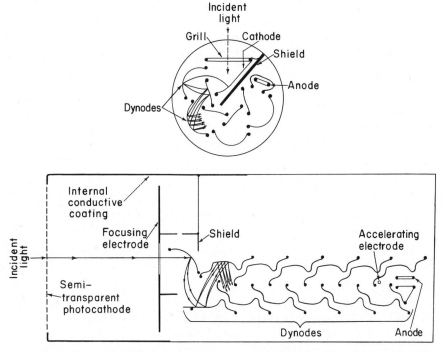

Fig. 113. Internal structure of RCA photomultiplier tubes: (*upper*) 1P21 and (*lower*) 6810. (Courtesy Radio Corporation of America.)

mary electrons. The primary and secondary electrons can be brought to another metallic surface (dynode) which is at a higher positive potential than the previous one. Again, additional electrons are set free. This process may be repeated several times more until the electrons leave the cell through the anode. A 1P21 with nine successive stages thus can enormously multiply the relatively small number of primary electrons (Fig. 114).

The advent of the photomultiplier has made both gas-filled and vacuum cells largely obsolete for work in the ultraviolet and the

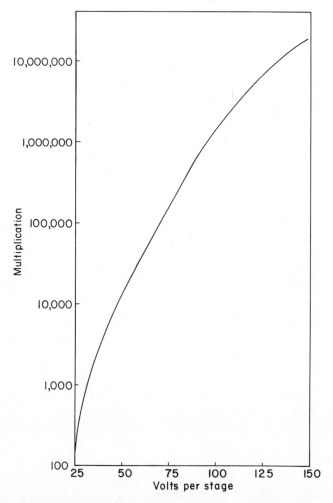

Fig. 114.   Multiplication of primary photocurrent of a photomultiplier tube (RCA (1P21) as depending on voltage per dynode stage.

visible region of the spectrum (Fig. 115). But such cells are still used in the red and infrared.

The 1P21 and other multipliers can be used with a galvanometer in the output circuit without any auxiliary circuit elements. However, with the very convenient recording meters now available, it is preferable to have the multiplier followed by an electronic amplifier, usually with provisions to compensate for the dark current of the cell, and with a gain that can be adjusted in order that the deflection of the recording meter may be matched to the width of its strip chart.

A complete photoelectric photometer consists of three major components: the photometer head with the detector at the telescope, the amplifier, and the recording meter (Fig. 116).

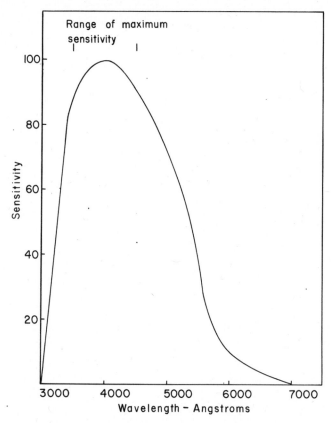

Fig. 115.   Spectral response of 1P21 photomultiplier tube; sensitivity in arbitrary units.

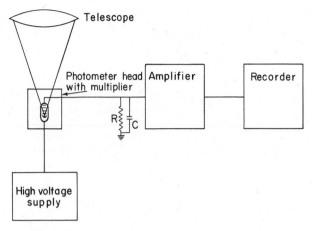

Fig. 116.   Scheme of a photoelectric star photometer.

A fairly high voltage is required to make the potential of the dynodes in a photomultiplier consecutively higher. The potential difference between the cathode and the last dynode of most multipliers now in use for stellar photometry varies from 700 to about 2000 volts. The current multiplication within the multiplier varies very sensitively with the voltage across the dynodes. Expressed in percentages for small voltage changes, the variation of multiplication is proportional to the voltage change multiplied by the number of dynodes. Obviously, even a small instability of this voltage, say of the order of 1 percent, causes intolerably high variations of current multiplication, with a resulting low accuracy of the measurements.

The dynode voltage can be obtained from batteries, which will last for considerable time since almost no current is drawn. On the other hand, batteries supplying such high voltages represent a considerable hazard to the observer. Highly stabilized electronic power supplies are therefore frequently preferred. They consist of a rectifier stage operated on the power line and a stabilizing stage to compensate line-voltage fluctuations.

The high-voltage supply is an essential part of the photoelectric photometer, and great care must be taken to insure the utmost stability.

The photometer head (Fig. 117) containing the photomultiplier is the unit that is attached to the telescope. The multiplier is frequently surrounded by a chamber in which a cooling medium such

Fig. 117.   Head of a photoelectric star photometer. (Courtesy Lowell Observatory.)

as Dry Ice can be placed for refrigeration of the cell. Refrigeration reduces the dark current of a photocell; if one is to be used down to its very limit, cooling is essential. On the other hand, cooling may easily create undesirable side effects such as water condensation and ice formation. This, in turn, may lead to current leakages and other operational difficulties.

Close to the multiplier is a lens, the so-called Fabry lens, which focuses the exit pupil of the telescope onto the photocathode. The image cast on the cathode is an image of the objective lens illuminated by the star. If there is unsatisfactory guiding of the telescope or poor seeing, this image does not move on the cathode. The photosensitive layer of the cathode is never perfectly uniform, and a slight motion of the image over the cathode might vary the photocurrent.

A small diaphragm or a set of interchangeable diaphragms with different diameters is located close to the Fabry lens in the focal plane of the telescope. Diaphragms of different sizes are used to accommodate the star image under different seeing conditions. The one being used cuts off the background radiation of the sky except for that immediately surrounding the star. In regions crowded with stars, this diaphragm also prevents the light from neighboring stars from reaching the photocathode.

Before the light strikes the cathode it passes through color filters which are provided to select the wavelength region in which the measurements are taken.

Frequently photoelectric photometers are equipped with a source of constant brightness made of a radioactive paint. The radiation of this source can be used as a standard to check the over-all sensitivity and stability of the photometer, including its electronic components.

An eyepiece is provided for setting and guiding on the star. It is wise to arrange this eyepiece so that the stellar image can be seen in the diaphragm to assure proper setting. The possibility of seeing a fairly large field in this eyepiece if desired or in a second eyepiece is useful for identification.

Two sources of light contribute to the illumination of the photocell: the light of the star to be measured and the light of the sky background. The fainter are the stars under observation, the larger becomes the percentage which the sky contributes to the output of the photocell. Eventually the sky contribution will become larger than that from the star; the equipment reaches its limit. Since the observer desires to know the brightness of the star alone, the measurements have to be corrected for the background light. To find the amount of this correction, the observer measures the background light that passes through the diaphragm from the vicinity of the star (Fig. 118). This can be done simply by moving the telescope off the star to a region free of other stars, but it is preferable to provide an offset mechanism for the photometer which permits swinging it on and off the star without moving the telescope. Such an offset mechanism in two perpendicular axes is also very convenient if the photometer is to be used on objects too faint to be visible to the eye through the guiding eyepiece. Then the observer needs to know only the angular distance of the object from a nearby visible star. The telescope is pointed to this brighter star, and the offset mechanism is manipulated to move the photometer from the reference star by a predetermined amount.

The output of the photomultiplier is ordinarily a direct current. Occasionally, an alternating photocurrent is preferred. A suitable device for the modulation of the beam has to be provided. A mechanical chopper or a rotating polarizing element serves the purpose.

The photocurrent is led to the amplifier by a shielded cable. The same minute attention has to be paid to the input stage of the amplifier as to the photomultiplier. Creeping currents at spots of poor insulation or even breakdowns of insulation, perhaps caused by humidity, must be prevented by all means. With the multiplier, the input impedance of the amplifier might be as high as several hundred megohms. If the photocurrent has to be measured within 0.1 percent or better, the insulation of cable, tube sockets, and other parts ought to be higher than 10^{11} ohms. Care must be exercised in the use of coaxial cables and movable contacts; sometimes they generate spurious signals.

The design of the amplifier depends in the first place on whether

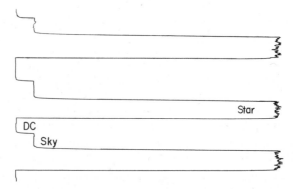

Fig. 118.   Record produced by a photoelectric star photometer.

the photometer puts out direct or alternating current, or whether one wants to count photons impinging on the photocathode. In a photoelectric photometer employing a dc amplifier, the photo current produces a voltage across a resistor R, and this voltage is applied to the control grid of the input tube. It is wise to use resistors with low temperature coefficients to reduce the sensitivity changes of the amplifier as much as possible. The resistor and a condenser C across it determine the time interval necessary to reach full deflection of the meter following the amplifier. This interval depends on the product RC, which is called the time constant of the amplifier. It is ordinarily chosen to be of the order of 1–4 seconds. The larger the time constant the more the amplifier smooths out the rapid variations in the photocurrent. This "noise" arises from various sources, including atmospheric seeing. The noise in the amplifier output will be more suppressed, if the time constant is large, because the amplifier cannot follow these fast current changes. On the other hand, with a large time constant the observer needs to wait considerably longer before full deflection is reached and a reading can be taken.

An extremely desirable feature of the amplifier is a linear relation—that is, direct proportionality—between photocurrent and amplifier output. The aim should be to keep deviations from linearity smaller than 0.1 percent. This can be achieved by a large negative feedback of the output into the input of the amplifier (Fig. 119). The feedback reduces the gain, but it also tends to linearize the performance of the amplifier. The feedback can be made so large

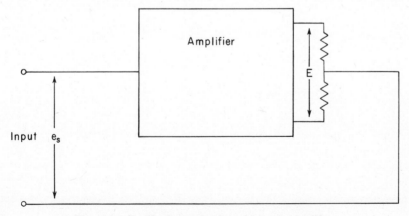

Fig. 119.   Feedback circuit of an electronic amplifier.

that the ratio of output to input signal becomes almost independent of the gain of the amplifier without the feedback circuit, and thus independent of variations of characteristics of individual tubes.

A detrimental property of a dc-amplifier is sometimes the instability of its output when no signal is applied. This "zero drift" can be rather annoying, and if it is rapid and erratic it affects the accuracy obtainable for the measurements. A great deal of drift is often caused by changes of supply voltages. Bias and plate voltages can be kept stable within the requirements. The critical factor is the filament voltage of the first stage of the amplifier. For an amplifier of 0–0.1-volt input range, as usually used in photoelectric work, the filament voltage must not vary more than 0.1 percent. Otherwise, the plate current would vary by an intolerable amount. This sort of constancy is hard to establish. The difficulty of the problem is noticeably reduced by an amplifier design employing a differential circuit (Fig. 120). If the two tubes are exactly identical in their characteristic output, voltage across Z would be independent of the supply voltages of the tubes, and of the filament current. It is advisable to select for the input stage two tubes whose characteristics are as nearly equal as possible. For small changes of the operating point on the characteristic curve such a circuit still remains in balance. Short-term drifts remain negligible, even with changes in filament current of the order of 1 percent. With aging of the tubes, differential effects may become serious and the circuit may need rebalancing

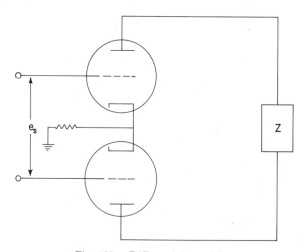

Fig. 120.   Differential amplifier.

or a new pair of tubes may have to be selected. Differential amplifiers with matched tubes are also advantageous in so far as they yield
increased linearity.

Alternating-current techniques are sometimes used in stellar photometry with photoemissive cells for the special nature of the problems to be solved. Occasionally, they have been preferred in the past
by observers because of the greater stability of the ac amplifier. This
is barely true now, since we have learned to design dc amplifier
circuits that are equally stable.

Alternating-current methods have found wide application in the
field of stellar photometry for the measurement of the polarization
of starlight, measurement of color indices without having to measure individually the magnitudes of the stars in two different wavelength regions, the continuous comparison of the brightness of two
stars with small angular distance, and in other special problems.

If the ac technique is to be applied to measure the intensity of
stellar light in the fashion of dc methods, a chopper—either a mechanical one, such as a rotating diaphragm, or an optical one, such
as a rotating polarizer—must be provided in front of the photocell.
The ac component of the photocell output is applied to the control
grid of an ac amplifier tuned for the chopping frequency. This
amplifier is followed by a rectifier to convert the amplified signal
into a dc signal to be recorded by a meter. It is desirable to have a
strictly linear relation between the ac signal of the cell and the dc
output of the amplifier. Fundamentally, rectifying tubes or dry rec-
tifiers are nonlinear devices with curved characteristics, particularly
at low voltages. This drawback can be overcome by amplifying the
ac signal voltage to a large value and then rectifying it. Negative
feedback is also commonly used to improve linearity.

With ac photometers null methods have been used for detection.
They have some advantages, such as relatively uncritical demands
on the electronic circuits. Öhman's photometer for measuring color
indices serves to illustrate the principle. The modulator is a rotating
cone consisting of six occulting and six transparent sectors located in
the beam of the light coming from the objective of the telescope
(Fig. 121). The occulting sectors are painted white on the lower surface in order to give diffuse reflection from a comparison lamp at
the side. A diaphragm is arranged in front of the phototube so as
to define a solid angle of the diffused light equal to the solid angle
subtended by the objective of the telescope. The angle of incidence

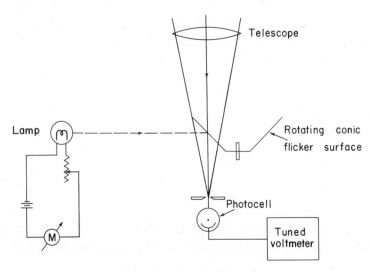

Fig. 121.   Flicker photometer of Öhman.

is nearly constant all over the small illuminated area that produces the comparison light when the conical surface rotates.

In this way, the starlight is periodically replaced by the comparison light. By varying the brightness of the lamp, it is possible to make the flicker current almost disappear and accurate settings on minimum ac current can be obtained. The brightness of the lamp can be changed by changing the current through it. The current is used as a measure of the brightness of the star. It is also possible to keep the current constant and vary the comparison light with a wedge until balance is established.

Walraven has developed a photometer that combines the null method with a servo-mechanism system permitting measurement of the ratio of intensity of two stars of small angular distance. The two beams are alternately directed onto the cathode of the phototube. The resulting ac signal is used to control an equalizer (for example, a wedge) until equilibrium is reached. The equilibrium position of the equalizer represents the brightness ratio. The linearity and stability of the amplifier are unimportant features. Walraven's arrangement has the remarkable property that it looks at the two objects simultaneously. The instrument has a very outstanding advantage for variable-star work. Continuous light curves can be obtained in a highly automatic fashion provided the guiding of the telescope is satisfactory. Another advantage is the fact that changes of atmospheric transmission are much less harmful than with other methods because of the continuous reference to the comparison star.

In addition to photoelectric star photometers employing dc or ac techniques, a third kind has been developed which couples a multiplier with a pulse-counting apparatus. Pulse-counting photometers such as the one used by Baum with the 200-inch Palomar Mountain telescope have been particularly successful in work on very faint objects (Figs. 122 and 123).

Under the impact of radiation, electrons are freed from the cathode of a multiplier and converted into avalanches of electrons by the multiplying action of the secondary-emission dynodes. Thus the output current is not steady but consists of a sequence of pulses. By counting the pulses produced within a selected interval of time a measure of the intensity of the light falling on the cathode can be obtained. The pulses are amplified in a pulse amplifier. Their amplitudes differ considerably, depending on whether the pulses are due to electrons released at the photocathode or to electrons originating at later stages without radiation impact and contributing to the dark current.

Only the pulses due to the photoelectrons are of interest. A discriminator or "gate" eliminates pulses with less than a selected

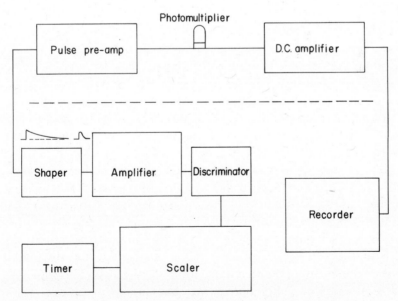

Fig. 122.   Block diagram of Baum's pulse-counting photometer.

Fig. 123.   Head-unit of Baum's pulse-counting photometer. (Courtesy Mount Wilson and Palomar Observatories.)

amplitude, whereby a considerable amount of dark current and noise is removed. The pulses permitted to pass through are fed into a binary scaler, which consists of a train of "flip-flop" halvers, each of which divides the number of pulses by two by passing only alternate ones. After the number of pulses has been reduced to the capacity of a mechanical counter, they are recorded. Sometimes, modifications of the counting arrangement allow counting in a scale of ten.

The time interval during which the pulses are counted is important in so far as it is required to be highly constant if consistent results are to be obtained. The accurate timing of the exposure is usually done automatically. Various devices are used as timers. Occasionally the power frequency of the line supplies the time standard and synchronizes a multivibrator pulse generator. To take a measurement the observer starts the binary scaler. The gate then closes the exposure after a preset time has elapsed, and the mechanical counter is read.

Pulse-counting systems can lack linearity as can other photoelectric techniques. If the system records pulses generated in the multiplier, linearity can be assumed between light intensity and pulse count, except at high counting rates, when pulses may get lost without being counted because they follow each other too rapidly to be resolved. Resolving times of the order of 1 microsecond are required at the light levels usually involved in stellar photometry.

For photometric measurements in the far red and infrared region of the spectrum no photoemissive cells or multipliers are commercially available which meet the severe requirements of stellar photometry, with its extremely low light levels. Fortunately, detectors of the photoconductive type sensitive to infrared radiation fill the gap. Though all semiconductors show evidence of photoconductivity, only a few are particularly useful in stellar photometry. The lead salts such as PbS and PbTe have been recognized as the most satisfactory detectors and probably will continue to remain so.

The scheme of an application of lead sulfide cells has been developed at the Harvard College Observatory (Figs. 124 and 125). A two-element cell is used, the star image being nutated alternately onto each element. The voltage output of each element of the cell consists of periodic pulses induced by the image motion. In addition, random noise voltages are present which are generated by the cell

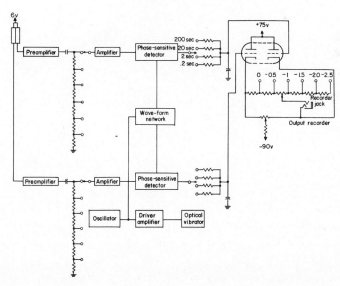

Fig. 124. Diagram of infrared star photometer of Harvard College Observatory.

Fig. 125.   Head of infrared star photometer of Harvard College Observatory; light comes in from below, the cell is in the upper right-hand corner, and the two Teflon plates (light colored) carry the preamplifiers.

without radiation. The voltages from the two elements are fed into separate amplifiers, one of which changes the sign of all the voltages that it passes.

After this initial amplification, the output voltages can be added. The resulting signal, consisting of a sine wave with the frequency of nutation and the random noise fluctuations, can now be amplified further by a single amplifier. A multivibrator generates a square wave at the frequency of nutation. It is triggered by the action of the nutating element in the photometer. The square wave is needed to reference a phase detector, into which the amplifier output is also fed. This detector passes unaltered all input voltages during the positive phase of the square wave and reverses the sign of all inputs during the negative phase. Thus the sine-wave component of the input which is also at the square-wave frequency will be rectified. The noise component, being random, will be observed in the output also as some random voltage. A charge proportional to the output but somewhat modulated by the noise is developed on a capacitor through an RC filter.

A remarkable feature of the system is that only the stellar image produces an ac signal while the illumination of the tiny cell elements by the background light of the area surrounding the star stays constant. The amplifier does not pass this dc signal.

Guiding of the telescope has to be very accurate to maintain steady positioning of the nutating stellar image on the minute cell elements. The system described also generates a guiding-error signal which is used to indicate the need for manual corrections to the telescope guidance or is fed back into an automatic mechanism correcting guide errors, in which case no visual monitoring is needed to secure proper guiding while the cell is exposed to stellar light.

Photoelectric Polarimetry

During the last few years photoelectric techniques have been employed successfully to detect and to measure polarization of starlight. The amount of polarization of stellar light is small and only the photoelectric methods are accurate enough to produce reliable results. The work is largely due to W. A. Hiltner and J. S. Hall, who approached the problem in somewhat different ways.

Hiltner's equipment consists of a regular photoelectric dc photometer. A piece of high-quality Polaroid placed in the beam close to the focus of the telescope serves as an analyzer. The Polaroid occupies either of two angular positions 90° apart with respect to the photometer. Photometer and Polaroid can be orientated as a whole on the optical axis of the telescope to any desired angle.

To determine the polarization of stellar light, the photometer and

Polaroid are set at some selected position angle. Three settings are read, the first and the third at the same position of the Polaroid and the second with the Polaroid rotated by 90°. The photometer and Polaroid are then set at another angle, 30° from the first and the three deflections are repeated. If the starlight is noticeably polarized, an estimate is made of the position angle of the plane of vibration. Then, one observation is made with the Polaroid approximately at one of its two positions in the plane of vibration. This observation determines the amount of polarization. The other two are made to determine with accuracy the position angle of the plane of vibration. This is done by setting the photometer at a position angle of 45° on either side of the estimated plane of vibration. The resulting observations are plotted with the difference in magnitude of the deflections when the Polaroid was rotated 90° against the position angle. The curves are then compared with a series of master sine curves, and the amount of polarization and the position angle can be determined by the master curve that best fits the plotted observations.

Hall's method of measuring polarization of stellar light involves an ac technique. A 1P21 photomultiplier produces an alternating current when struck by starlight that has passed through a Glan-Thompson prism [Fig. 10(b)] rotating at constant speed. Amplitude and phase of the current are the two quantities related to amount of polarization and position angle of the plane of vibration. The equipment can be arranged in five different conditions to obtain records of measurements with a recording meter (Fig. 126). In condition G, the light is collimated and allowed to strike the rotating prism without suffering sensible polarization in the optical train above it. If the starlight is polarized the photocurrent varies with the angle through which the prism is rotated beyond the position of maximum light transmission. With the aid of a phasing switch a square wave of frequency 30 cycles per second is generated. This switch is linked mechanically with the prism rotating at 15 revolutions per second and mixed with the signal in a synchronous amplifier. The relative phase of the two waves is changed completely every 2 minutes and the dc output voltage of the amplifier goes through a cyclical change shown on the chart recorder every 2 minutes.

In condition D a quartz depolarizer is substituted for the glass disk. The disk is inserted in the beam if the depolarizer is removed in order to make up for reflection and absorption losses in the de-

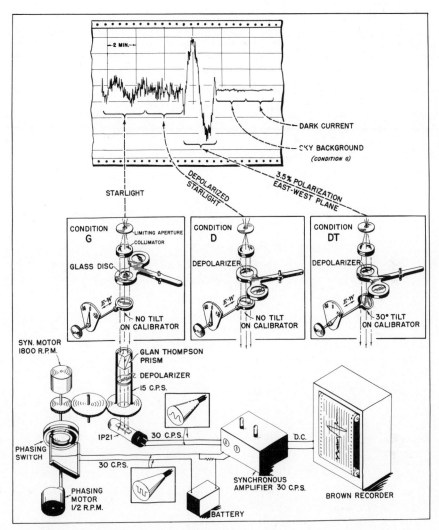

Fig. 126. Hall's photoelectric photometer for the measurement of polarization of stellar light. (Courtesy U. S. Naval Observatory.)

polarizer. The starlight is completely depolarized by the device before it reaches the analyzer.

The depolarizer is also used for condition *DT*. The cover glass that serves as a calibrator is tilted a certain amount, thereby introducing polarization in the transmitted starlight amounting to a fixed percentage. The axis of rotation of the cover glass lies in the plane of polarization of the transmitted light.

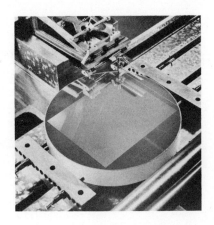

6

Spectroscopy

A wealth of information on the nature of stellar radiation and its sources has been obtained with the help of an instrument that permits highly detailed analysis of the light reaching the surface of the earth from cosmic objects. This tool, the spectroscope, is basically the same instrument that has accomplished tremendous achievements in a large variety of fields in physics, chemistry, medicine, and even mining and criminology.

The basic discovery leading to the conception of the spectroscope was made by Isaac Newton in 1666. He found that white light is made up of components of different color. A glass prism was the means to spread the white light into its constituents.

The dispersing element of a spectroscope can be either a transparent prism or a diffraction grating. Newton produced the sun's spectrum on the wall of a dark room with the help of a prism, but it was not until W. H. Wollaston (1802) and J. Fraunhofer (1814) introduced a narrow slit in front of the prism that spectra with resolved features could be obtained. G. R. Kirchhoff and R. Bunsen constructed in 1859 the first practical spectroscope. Since then, spectroscopy has made tremendous progress and has solved a long list of problems in all fields of science. Astrophysics has become possible only because of spectroscopy; in fact, it has been at times mainly spectroscopy applied to cosmic light sources. It was spectroscopy that disproved so radically the famous dictum of Auguste Comte (1798–1857): "There are some things of which the human race must remain forever in ignorance, for example, the chemical constitution of the heavenly bodies."

Prisms and Gratings

The basic design of astronomical spectrographs does not differ very much from that of instruments for use in the laboratory. Astronomy requires numerous modifications owing to the particular nature of the observations with a telescope and to the operating conditions, which are different from those encountered in a laboratory. As in laboratory instruments, prisms or gratings are used as a means of dispersion. Prisms have been preferred in the past, but diffraction gratings have become more and more common in recent constructions, particularly since we have learned to increase their efficiency by properly shaping the grooves.

The dispersive power of a prism arises from the variation of the refractive index of the prism material with wavelength (Fig. 127). Rays of short wavelength are refracted more on passing through a prism than rays of longer wavelength. The variation of the refractive index depends on properties of the material used for making the prism. The dispersion is not uniform for all wavelengths; the red end of the spectrum is usually considerably more crowded together than the blue end. The dispersion also depends on the refracting angle. Refracting angles of 60° are common. Larger angles

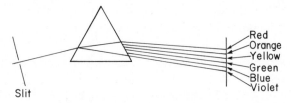

Fig. 127.   Dispersion of white light by a prism.

would give larger dispersion, but because of the longer light path through the prism larger light losses would occur by absorption.

Ordinarily the shape of a prism permits a beam of circular cross section to pass through it at the angle of minimum deviation, that is, so that the angle between the incident and emergent beams is minimum. For a 60° prism this requires a ratio of the length to the height of the prism of about 1.6.

An important property of the prism or the diffraction grating is the resolving power. It is the wavelength interval between the closest spectral lines of similar intensity that can be distinguished in the spectrum. The resolving power of a prism increases with the linear width of the beam entering the prism.

Choice of the prism material is made not only with respect to the dispersion desired but also with consideration for the spectral region to be investigated. Various optical materials differ both in their dispersive power and in their ranges of transmission. Most glasses are quite dispersive and transparent in the visible region of the spectrum. Crown glasses are more transparent in the violet and ultraviolet than the heavier flint glasses, but they are somewhat less dispersive. Very light crown glasses (ultraviolet glass) are therefore used in astronomical spectrographs intended to reach the ultraviolet spectral region.

Still more transparent in the ultraviolet is quartz (SiO_2), but the dispersive power is less, especially in the visible and in the red. Quartz is available in two modifications for optical purposes. Natural quartz comes in crystals. Pieces satisfactory for large prisms are hard to obtain and expensive. Crystal quartz is birefringent, so that an incident beam is split into two refracted beams; however, the formation of double images can be prevented, more or less, by cutting the quartz in such a fashion that the optical axis of the prism coincides with the optical axis of the crystal. A slight doubling sometimes remains. Doubling of the spectrum can be eliminated completely by a method introduced by Cornu. Half of the prism is cut from a crystal producing right-handed rotation and half from a crystal producing left-handed rotation of the plane of polarization (Fig. 128). The two halves are cemented together in such a manner that the two effects compensate each other, and a single image is obtained.

Techniques now make it possible to fabricate lenses and prisms from fused or vitreous quartz which does not display birefringence

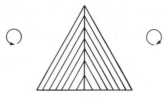

Fig. 128.   Cornu prism.

or circular polarization. Vitreous quartz is less dispersive than crystalline quartz and it is also less transparent for wavelengths shorter than 2800 A. This matters very little for ordinary astronomical applications because the spectral region in question is inaccessible from the surface of the earth, owing to the high opacity of the atmosphere.

Rock salt (NaCl) and potassium bromide are transparent over a large region of the spectrum extending far into the ultraviolet and the infrared. These materials are rarely used in stellar spectrographs because they are highly hygroscopic and would be easily destroyed by atmospheric moisture. Effective use of rock salt and potassium bromide would require careful precautions. The working conditions to which a stellar spectrograph is exposed make such precautions difficult.

Fluorite (calcium fluoride) is an optical material that is transparent from the infrared to about 1250 A; its dispersion in the visible region is small. Fairly large crystals can be grown now, but big pieces are still expensive. Lithium fluoride is transparent even to 1050 A, but, because of its brittleness, it is difficult to work with. Use of these two materials is advisable only if one wants to obtain information on the far ultraviolet with the aid of rockets to carry the equipment beyond the absorbing layers of the atmosphere.

Dispersion and resolving power can be increased by sending the beam of light through several prisms in a train or by making it pass the same prism twice. Many two- or three-prism spectrographs are in use for stellar work (Fig. 129). Multiprism spectrographs are of low efficiency because of the light losses by reflection at the many surfaces of the optical parts and by absorption and scattering in the large amount of glass the light has to pass. Modern diffraction gratings, particularly those of the blazed type (defined below), are becoming more common in astronomical spectrographs.

The dispersing action of diffraction gratings has been known since the time of Fraunhofer (1821). Such gratings have been widely used in laboratory spectroscopy, but it was only fairly recently that the difficulties of developing the potentialities and reproducing good gratings in sufficient quantities have been overcome.

A diffraction grating consists of a large number of narrow, straight lines ruled parallel and equally spaced on a suitable reflecting or transmitting surface. The gratings are called, respectively, a reflection or a transmission diffraction grating. The number of lines on a grating can be rather high. A grating with 20 lines per millimeter is a very coarse grating; 1200 lines per millimeter might be ruled to obtain a very fine grating. The more closely the lines are placed together, the greater is the dispersing power of the grating. The total number of lines on the grating determines its resolving power, that is, its ability to separate two spectral lines with a very small wavelength difference existing between them.

The making of a grating is a delicate and difficult procedure. Fraunhofer and others, among them Rutherford in 1870, tried to produce small good gratings. Rowland at Johns Hopkins University, between 1882 and 1901, was the first who succeeded in making high-quality gratings, on concave as well as on plane surfaces. Even now, only a few people are able to rule high-quality gratings.

The grooves are ruled with a properly shaped diamond. The quality of the grating depends mainly on the accuracy that can be attained in spacing the grooves uniformly. The heart of the grating-ruling engine (Figs. 130 and 131) is the precision screw, which moves the diamond perpendicularly to the direction of the grooves each time one groove has been ruled, and its bearings. To accomplish best results, the nature of the errors of the screw should be known. Small errors will always exist and will be detrimental enough to prevent the ruling of large gratings. An electronic device

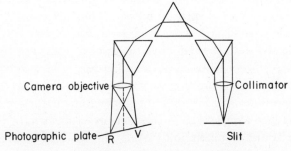

Fig. 129.   Diagram of a three-prism spectrograph.

Fig. 130.   Grating-ruling engine. (Courtesy Bausch & Lomb.)

has been designed that uses the wavelength of a sharp spectral emission line as the standard of length to plot the error curve of the screw. This curve supplies the data necessary for cutting a metal cam that produces the exact corrections needed to compensate for the screw errors.

A diffraction grating does not disperse the light only over a single spectrum. It produces an undispersed image of the source (or of the spectrograph slit), which is usually referred to as the zero-order spectrum. On both sides of this zero-order spectrum in the direction of dispersion a series of spectra appear. They are called spectra of the first order, the second order, and so on (Fig. 132). The blue light is less deflected than the red—just the opposite of prism spectra.

In the first-order spectrum the angular deviation of light is nearly proportional to wavelength, a very advantageous feature for wave-length determinations. The amount of spreading in the second-order spectra is twice as much as in the first-order spectra, in the third-order three times as much, and so on. This leads to an overlapping of different orders; for example, the red of the first order overlaps the blue of the second. By inserting suitable color filters and by the proper selection of photographic emulsions, the confusion can be avoided to some extent. However, the total available light is dissipated over a large number of spectra, when usually only one is actually desired. This waste of light is a rather serious disadvantage for the application of gratings in an astronomical spectrograph to such faint light sources as the stars.

Light losses by production of many spectra of different orders as well as disturbing overlapping have been considerably reduced by a technique of ruling gratings called blazing. In ruling modern diffraction gratings, the ruling diamond is lapped and polished to such a shape that the faces of the grooves it rules will have the required blaze angle (Fig. 133). This angle is determined by calculating, from the wavelength of the center of the desired region of the spectrum, the angle at which the diffracted light will leave the grating; the blaze angle is chosen so that the light reflected from the groove face as from a mirror will leave the grating at the same angle as the dif-

Fig. 131.   A grating on the ruling engine. (Courtesy Bausch & Lomb.)

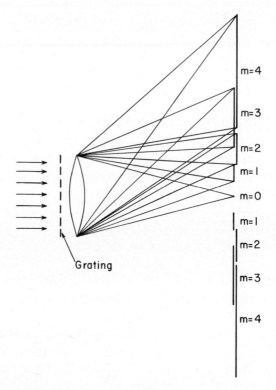

m=4

m=3

m=2

m=1

m=0

m=1

m=2

m=3

m=4

Grating

Fig. 132.   Overlapping of grating spectra of different orders m for a given grating constant and a wavelength range of 3000–6000 A.

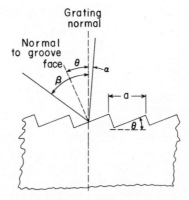

Grating
normal

Normal
to groove
face θ

α

β

a

θ

Fig. 133.   Cross section of a diffraction grating: a, width of groove; α, angle of incidence, β, angle of diffraction for blazed wavelength; θ, blaze angle.

fracted light. By this means most of the incident light is concentrated into a particular order and wavelength range of the spectrum. The increase in speed and efficiency of astronomical spectrographs with modern blazed gratings is tremendous. High-dispersion spectrographs can now be used for investigating faint stars that were completely out of reach with spectrographs of the same dispersion but with ordinary gratings or prisms as dispersing elements.

Until recently the gratings employed in spectrographs were original gratings as ruled by the engine. Ruling a grating is a slow process, and a ruling engine might turn out only a few per year. This makes good gratings expensive and hard to get. Since 1947, techniques for producing replicas from master gratings have been so successfully improved that now first-class diffraction gratings are easily obtainable in quantities. More than 50 years ago, a process was devised by T. Thorp and improved by R. J. Wallace by which replicas could be made. A solution of collodion or gelatin was poured on the surface of a grating. After this coat had hardened, it was stripped off and mounted on a plate of optical glass. Only transmission gratings could be produced in this fashion, and part of the resolving power of the master grating got lost in the replica.

In 1947, White and Frazer of the Perkin Elmer Corporation developed a technique of producing better replicas by making a thin plastic model of a master grating, coating it with an evaporated aluminum film, and backing it with a flexible glass plate. The grooves are formed in the very thin layer of clear plastic that adheres to the surface of the glass backing plate. This process, or a similar one, is now used by various commercial manufacturers. The coating with aluminum is required only if a reflection grating is wanted; transmission gratings, of course, do not need coating. A large number of replicas can be made from a single master grating.

Objective Prisms and Gratings

In ordinary spectrographs a collimator is provided to make parallel the light striking the prism or grating. A slit illuminated by the light to be investigated acts as the source (Fig. 138). In the case of stars, the light is already parallel before it falls upon the objective of the telescope. If a prism (Fig. 134) or a grating is placed in front of the objective we have a simple form of stellar spectrograph.

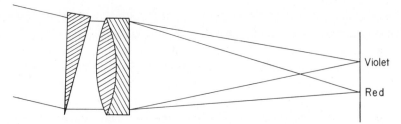

Fig. 134. Objective prism.

The light is dispersed by the prism or grating and then brought to an image as a spectrum in the focal plane of the telescope.

Such an arrangement was used by Fraunhofer as early as 1823, but lost importance for a long time by the introduction of the slit spectrograph. In recent times, a number of useful applications have been found. This arrangement has advantages that make it superior to the slit spectrograph for certain purposes. The most outstanding advantage is its ability to produce spectra of a large number of stars simultaneously over the field of the telescope imaged on the photographic plate (Fig. 135).

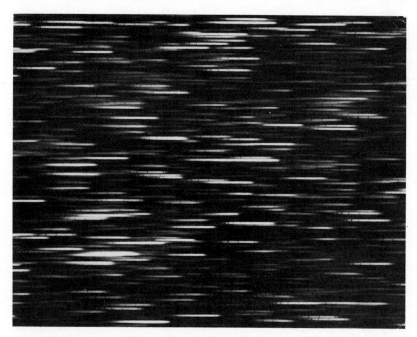

Fig. 135.   Objective-prism spectra. (Courtesy C. Fehrenbach.)

The focal lengths of large telescopes are considerably longer than the focal lengths of cameras usually employed in stellar slit spectrographs. Therefore, if prisms are used with refracting angles as large as in ordinary spectrographs, the linear dispersion of the spectra will be high, but the refracting angle of objective prisms can be kept small and the dispersion will still be sufficient. Small prism angles mean also that the glass through which the light has to pass is not particularly thick and light losses by absorption in the glass will be kept small.

The objective prism is a powerful instrument as far as speed is concerned. Yet the arrangement has a number of disadvantages. The resolution of slit spectra depends upon various factors, among them the width of the spectrograph slit. In the case of the objective prism, the slit width is replaced by the diffraction and seeing disk of the star. With a slit spectrograph, a large seeing disk results in a loss of light on the slit jaws. Only a small fraction of the total available light can enter the narrow slit, but the quality of the spectrum will not be affected by the size of the seeing disk. The situation is different with the objective prism. Poor seeing (resulting in a large image disk) causes a loss of resolution, and fine details in the spectrum do not show. In a slit spectrum, the spectral lines are images of the slit produced by nearly monochromatic light; in an objective-prism spectrum, the lines are images of the seeing disk.

Another drawback of the objective-prism arrangement lies in the difficulty of putting a comparison spectrum with lines of known wavelength alongside the stellar spectrum for the purpose of determining wavelengths in it. It can be done by impressing on the spectra of the stars absorption lines of known wavelengths by means of a filter inserted in the beam. A suitable filter is a solution of neodymium chloride. This produces diffuse absorption bands in the yellow and red portions of the spectrum, which are useless for our purpose, but in the blue a fairly sharp band occurs which can serve as a wavelength standard.

The neodymium absorption does not satisfactorily meet all the requirements to be desired from a wavelength standard. It seems that the exact wavelength of the band varies somewhat with the concentration of the solution and its temperature. Also, the distribution of intensity across the band is not symmetric; the position at which its center seems to be located and on which the observer would set the wire of the eyepiece of the measuring engine depends upon the density of the photographic plate. Errors, hard to control, are introduced by these properties of the band.

One use to which wavelength measurements can be put is the determination of radial velocities of stars, that is, the velocity component along the line of sight. When a star is moving away from the observer, the wavelengths of its spectral lines are increased, that is, the lines are shifted toward the red end of the spectrum (the Doppler effect). If the star is approaching the observer, the shift is in the opposite direction. The change in wavelength depends on the radial velocity.

R. W. Wood suggested a technique for measuring radial velocities by producing two spectra of the stars simultaneously. Two prisms, with parallel bases and the same prism angle but opposite orientation, are put in front of the objective. Most of the dispersion is provided by transmission gratings ruled on the prisms. Their first-order spectra would be far apart on the plate without the prisms, but the deviation produced by the prisms brings the two spectra side by side.

There are several modifications of this method. In one of them two zero-deviation prisms are employed. A zero-deviation or direct-vision prism is a combination of two or more prisms of different refractive index, chosen in such a way that light of a particular wavelength passes through the combination without deviation, whereas light of other wavelengths is bent. If one such prism is used for each half of the aperture, the two spectra of each star can be produced side by side. The relative distances of corresponding lines of the twin spectra determine the radial velocities. The measured shifts lead only to relative radial velocities of the stars in the field of plate. If the true radial velocities of some of these stars are known, the relative radial velocities can be transformed into absolute velocities for all measured stars.

Results of this method and its modifications are subject to systematic errors on account of the large distortion of field produced by ordinary prisms. A noticeable improvement has been made by Fehrenbach; he proposed a combination of three prisms of glass of different refractive index (Fig. 136). Refracting angles and glass can be chosen in such a way that no field distortion results. The prism combination acts like a zero-deviation prism for a certain wavelength. The accuracy of radial-velocity measurements by Wood's method employing such a prism combination is remarkably good.

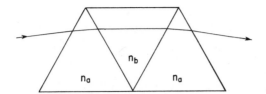

Fig. 136.   Fehrenbach's objective prism for radial-velocity determinations.

Gratings in front of the telescope objective are used very little for spectroscopic work for obvious reasons. Gratings produce several spectra of each star, thus considerably reducing the light available to go into a single spectrum of a certain order. Long exposure times result. Furthermore, photographs of stellar fields taken with objective prisms become rather crowded if they reach faint stars; individual spectra frequently overlap each other (Fig. 135). This is much worse if the photographs are taken with a dispersing device such as a grating that produces several spectra of each object in the field.

The use of objective gratings thus has been restricted to rather special problems, such as the setting up of magnitude scales in photographic photometry. The spectral images of different order have a definite magnitude difference depending on the geometric properties of the grating, and this magnitude difference is sometimes used as a yardstick for constructing photometric scales.

Slitless Spectrographs

Objective prisms for larger telescopes are not feasible because of the difficulties of manufacture. An optical arrangement somewhat similar in function to an objective prism, but with a fairly small prism, overcomes the difficulties and yet makes full use of a big telescope objective. The objective of the telescope, mirror or lens, serves simply as a light collector. The convergent beam coming from the objective is brought into a collimator employing a negative lens or mirror. It intercepts the beam before it reaches the focus and makes the light parallel again. The diameter of the beam of parallel rays is now considerably smaller than before it struck the objective. The light then passes through a prism or grating before it enters a camera.

If the telescope is a reflector and if the camera employs a mirror as an objective, the optical system is free of chromatic aberration. This is true also if collimator and camera lens are of the same focal length and type of glass; the two lenses combined without the prism have zero net power and act like a plane-parallel plate of glass put into the convergent beam. The spherical aberration is small enough not to matter, if the thickness of the plate (or the lenses) is moderate. If the dispersing prism is placed in the beam in such a way that the bending of the rays at each surface is the same, the aberrations introduced by the lenses are nearly canceled. A spectrograph of this type requires only simple and inexpensive optics.

If a slitless spectrograph is properly designed, it utilizes the full aperture of the telescope objective. Furthermore, if the collimator lens has a diameter large enough to accommodate not only the converging beam to the center of the field but off-axis rays as well, the usable field at the photographic plate is about the same as that of the telescope itself.

The collimator may employ a positive element instead of a negative lens. A slit spectrograph can thus be turned into a slitless one by removing its slit. The disadvantage of this system is its length. A negative collimator lens shortens the length of the apparatus appreciably, a favorable circumstance, because this leads easily to increased mechanical rigidity of the arrangement.

The quality of the spectra produced by a slitless spectrograph suffers from the finite diameter of the seeing disk of the stellar image, much as the objective-prism spectra do. Guiding errors and differential flexure of the telescope caused by mechanical imperfection of the instrument also may lower the quality of the spectra obtained. The large focal length of big telescopes produces seeing images of considerable linear diameter which destroy the definition of spectral lines. To make up for an increase in focal length the dispersion has to be raised approximately in proportion to the increase in focal length. If the f-number of the telescope is unchanged, the light-gathering power of the objective increases with the square of the diameter of the objective. For a given resolution and width of spectrum, the illumination at the place on the photographic plate where the spectrum is produced increases only with the linear aperture of the telescope.

The objective prism and the slitless spectrograph are useful and efficient instruments where the highest possible resolution of details of the spectra is not the aim. For classification work on stellar spectra, such as the determination of spectral types and spectral luminosity classes, the slitless arrangement is indispensable if large numbers of stars are considered. It is also the most suitable instrument if the energy distribution of the continuum of stellar spectra is to be measured. A slit spectrograph might easily lead to erroneous results. The stellar image projected on the slit might be affected by chromatic aberration of the optics of the telescope, or by the dispersing action of the atmosphere, particularly if the star happens to be observed at low altitude. In other words, different portions of the tiny stellar image in the focal plane might be illuminated by light of somewhat different composition of colors. Since the slit does not permit all the light of the image to enter the spectrograph, the intensity distribution over the length of the spectrum at the photographic plate might not be the true one but one subject to uncontrollable errors.

Wavelength and radial-velocity measurements on spectra taken with slitless spectrographs encounter the same difficulties as in the case of the objective-prism spectra. In principle, the methods suitable for the objective prism can be used with the slitless spectrographs.

Recently, P. J. Treanor devised a method of measuring radial velocities for a temperature-controlled slitless spectrograph. Temperature control of prism spectrographs is essential, because it makes sure that the dispersive power of the optics does not change during the exposure of the photographic plate. A zero-deviation prism is introduced into the collimated beam of the spectrograph (Fig. 137). This prism is so designed that it completely eliminates field distortion for the undeviated wavelength, which is determined only by the refractive indices n_1 and n_2 of the components of the prisms. It consists of two right-angled prisms, geometrically identical in construction, and a 60° prism in contact with the hypote-

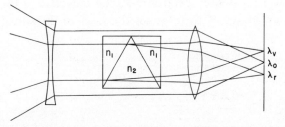

Fig. 137. Treanor's stellar image method for determining radial-velocities with a slitless spectrograph; λ_o, star image at position of undeviated wavelength; λ_v, λ_r, violet and red ends of spectrum.

nuse faces of the 90° prisms. The common principal section of the combination is rectangular. The dispersion curves of the materials of the prisms have different slopes, and the wavelength at which they intersect defines the undeviated color. For light of this wavelength, the combination behaves as a homogeneous plane-parallel plate of glass; the light will pass through undeviated, whatever the angle of incidence. Field distortion by the prism is therefore completely prevented. The performance of this combination is very similar to that of Fehrenbach's compound objective prism.

Treanor's compound prism does not occupy the whole aperture of the beam, but leaves free an outer annulus. Undispersed light from the stars in the field may pass through this ring and produce a star image coincident with the undeviated wavelength in the spectra of the stars. For the measurement of radial velocities, the undispersed star image is treated as a spectral line of corresponding wavelength, and is used in the same way as, for example, the artificial neodymium lines.

Slit Spectrographs

Spectrographic apparatus previously discussed is of specialized design and can be used only for stellar work of certain kinds. In general, a stellar spectrograph attached to a telescope consists of several major components: slit, collimator, dispersing means, and camera (Fig. 138).

The provision of a slit has two very important advantages. The resolving power can be increased and comparison spectra for measuring wavelengths can be produced on both sides of the stellar spectrum.

The slit, or part of it, is illuminated by the light source under investigation. In a stellar spectrograph the objective of the telescope produces an image of the star on the jaws of the slit. The size of

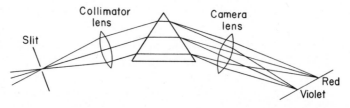

Fig. 138.   Basic prism spectrograph system.

the image depends on the seeing and on the diffraction disk of the star. With a telescope of long focal length the disk can be of noticeable linear diameter. If the slit is removed, the spectrograph produces monochromatic images of the seeing disk arranged according to wavelength at the focal plane of the camera. A large image disk destroys the fine details in the spectrum, because the large monochromatic images at the focal plane of the camera overlap each other. If a narrow slit is inserted at the focal plane of the telescope, only a part of the stellar image may be allowed to enter the spectrograph; the overlapping of the monochromatic images at the focal plane of the spectrograph camera will be less serious, and finer details can be resolved.

The location of the slit is at a fixed position relative to the rest of the instrument, including the photographic plate. If a portion of the slit is illuminated by the star, and two adjacent portions by an artificial light source, three spectra will be produced side by side (Fig. 139). Features having the same wavelength will coincide as far as their position on an axis in the direction of dispersion is concerned. By measuring the position of the stellar lines, relative to the position of the lines of the comparison spectra with known wavelength, the wavelengths of the stellar lines can be determined.

If the slit of a spectrograph is illuminated with diffuse light, it serves as the light source for the spectrograph. In the case of a stellar spectrograph the slit is the source only under certain conditions. It must be positioned at the focus of the collimator, and ought to be narrower than the stellar image. The aperture ratio of the collimator ought to be the same as that of the telescope or a little smaller; if it were larger, the starlight would pass through only the central portion of the collimator lens, and the diffuse light from the artificial source might fill the whole aperture. In case minor imperfections of the collimator optics exist, small shifts of star and comparison spectra are easily introduced. They would result in systematic errors of measured wavelengths.

Great care ought to be exercised in the construction of the slit. The accuracy with which it is made determines the quality of the spectra. The slit width should be variable. Two jaws forming the slit can be separated by a screw acting against a spring to make the jaws assume a well-defined position. In order to avoid a V-shape of the slit area, the edges of the jaws must be perfectly straight (except at the ends) and mounted truly parallel, and they

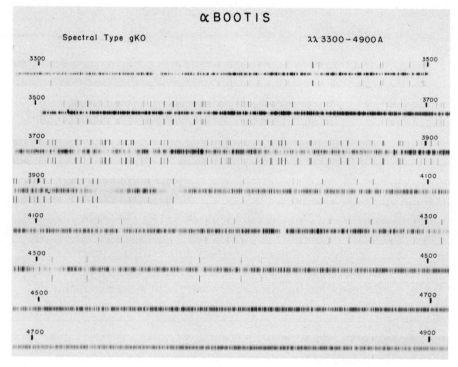

Fig. 139.   Spectrum of α Boötis, flanked by comparison spectra.

must remain parallel when they are moved to vary the width of the slit. The front faces of the jaws should lie in the same plane.

In a good slit, both jaws are moved simultaneously but in opposite directions to change the width. The center line of the slit thus does not move at all, regardless of the width adjustment. The screw driving the slit jaws should be calibrated so that the operator can adjust the width to a desired value. A slit should close completely only at its ends to avoid marring the sharp edges by touching each other.

The material of the jaws is usually a hard and durable metal, such as stainless steel. The adjacent edges forming the slit are beveled, so that light will not be reflected from them into the spectrograph. Ordinarily, the jaws are highly polished, so that their surfaces act like plane mirrors. Then part of the image of the star under investigation and the images of stars in its neighborhood can be seen on the polished jaws. The observer watches the luminous pattern on the slit with the help of the guiding eyepiece and makes sure that the telescope moves properly to keep the image of the star on the spectrograph slit.

The width of the spectrum perpendicular to the direction of dispersion on the photographic plate corresponds to the diameter of the stellar image at the spectrograph slit as projected through the spectrograph optics onto the plate. If the stellar image is small, the width of the spectrum produced at the photographic plate will be small; in fact, it might be too narrow to be suited for the purpose in mind. It may therefore be widened by making the stellar image trail back and forth over a suitably long section of the slit during the exposure. This trailing can be accomplished by introducing a slight drifting of the telescope. For producing the comparison spectrum, a diaphragm is put across the slit. It is cut so that no light from the artificial source can fall on the area of the stellar spectrum on the photographic plate, but the source light can reach a strip on each side of the stellar spectrum (Fig. 140).

The light leaving the slit is divergent. Prisms and gratings give greatest resolution if the rays illuminating them are parallel. The collimator serves to make the divergent rays from the slit parallel, and is expected to do so for rays of all colors. Fulfillment of this requirement is somewhat simplified, because all rays come from points of the slit close to the optical axis. Both spherical and chro-

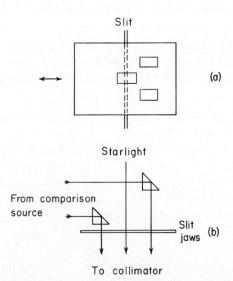

Fig. 140. Diaphragm and prism arrangement in front of slit for producing comparison spectra.

matic aberrations should be at a minimum in order to get a usable image of a long section of the spectrum.

If high resolving power is desired, large prisms or gratings are necessary and they must be fully illuminated. Hence, the parallel beam of light leaving the collimator ought to be of adequate diameter. This requires large collimator optics to accommodate the beam. The aperture ratio is given by the telescope. Thus, with telescopes of large aperture the length of the collimator, which is equal to the distance between the slit and the lens, becomes rather long. A collimator beam 100 mm in diameter would lead to a collimator length of 1.5 m with an aperture ratio of 1:15. The spectrograph thus becomes a bulky and unhandy piece of equipment to attach to the tube of the telescope. Also, such large equipment is difficult to construct so that it maintains perfect mechanical rigidity, no matter in what position it is moved with reference to the direction of gravity.

To reduce the size of astronomical spectrographs to be mounted on telescope tubes, folded collimators with mirror optics are often employed. Such a collimator might be like a Cassegrain reflecting telescope used "backward." The light from the slit enters through the central hole of the collimator's parabolic mirror. It is cast upon a small Cassegrain mirror and reflected back onto the large parabolic mirror which, in turn, reflects the light as parallel rays onto the dispersing prism or grating. Even large-aperture collimators operating with small f-numbers can be kept very short in this fashion and consequently they do not require a large mechanical layout for the spectrograph. Mirror systems have the additional advantage of not introducing chromatic aberrations.

Particularly in instruments with large beam diameter, gratings are preferred to prisms as dispersing elements. The light losses in large prisms become so great that they balance the advantages of wider beams. Since the advent of blazed gratings, spectrographs are more and more often equipped with them. Gratings are usually mounted on a turret so that they can be rotated in order to image the desired order or region of the spectrum on the photographic plate. More versatile instruments have facilities for swinging different gratings into the beam. By a suitable choice of the grating properties, such as the number of grooves per millimeter and the blaze angle, the optical spectrum can be well covered with different dispersions.

The camera of a stellar spectrograph should be designed to yield the utmost speed. The objective must accommodate the full width of the beam coming from the grating or the prism. The specific properties of the camera optics depend on the particular purpose of the instrument. Either lens or mirror optics can be used for the camera. Lenses as camera objectives do not necessarily need to be achromatic. By proper choice of the dispersion of the prism and the chromatic aberration of the lens, the foci of the camera for light of different wavelengths can be made to lie in a plane inclined to the axis of the camera; blue light is focused nearer to the lens than red light. The focal surface may be slightly curved, but a spectrum in good focus over a satisfactorily long wavelength range can still be obtained by bending the photographic plate so that its surface follows the curvature of the focal surface as closely as possible. Occasionally, film is employed instead of plates; bending of film is much less of a problem than bending of glass plates, but film is not advisable for wavelength measurements such as are required in radial-velocity work, though it is quite acceptable for photometric purposes. Spherical aberration of the camera optics for spectrographs should be corrected as perfectly as possible. If the curvature of the focal surface becomes too strong to be taken care of by bending of the photographic plate, an auxiliary optical element—a field flattener—is inserted in the beam in front of the plate. This is frequently done with fast Schmidt cameras.

The proper choice of the camera for a spectrograph depends on many factors, particularly on the dispersion desired. The linear dispersion increases with the focal length of the camera, and so does the exposure time necessary to take a spectrum of sufficient photographic density. The ratio of the width of the slit to the width of its image on the photographic plate is proportional to the ratio of the focal length of the collimator to the focal length of the camera. If this ratio is such that the slit is wider than its image, one might be able to open the slit rather wide to accommodate all of the light of the seeing disk of the star, and spectra might then be successfully taken even with poor seeing conditions. In a spectrograph whose camera has a large focal length compared with the focal length of the collimator the ratio of slit width to slit-image width is decreased, and the observer would want to keep the slit narrow. Since a large fraction of the total light available gets lost on the slit jaws, high-dispersion spectrographs are consequently of low efficiency under poor seeing conditions. Diffraction losses at the slit of an astronomical spectrograph rarely matter.

Modified microscope objectives with their small *f*-numbers have been used successfully for fast spectrograph cameras of low dispersion. The Rayton lens (Fig. 141) has become famous as a spectrograph camera objective with an *f*-number of 0.6. It has produced very important results in the radial-velocity work on galaxies. At 4350 A, spectra of objects as faint as of the 18th magnitude have been made, though with the very low dispersion of 440 A/mm and 510 A/mm. Oil-immersion microscope objectives with *f*-numbers of 0.35 are feasible for camera optics. Ordinary Schmidt cameras can be made with *f*-numbers up to 0.6. Still more efficient than the Schmidt in air optics are the solid-glass cameras of the Schmidt type (Chapter 3), with which *f*-numbers of 0.35 or slightly less can be made.

The actual design of a stellar spectrograph depends largely on the telescope with which it is used. A refractor has only one focus. In the case of a reflector, the spectrograph can be employed in various locations; it can be attached at the prime, the Newtonian, the Cassegrainian, or the coudé focus. Low-dispersion spectrographs are often attached to refractors. If they are used with a reflector they are attached to the prime or the Newtonian focus. They are the most effective types of instrument to take spectra for classification purposes or of objects that cannot be observed with higher dispersion because of their faintness. Reflectors usually have *f*-numbers between $f/3$ and $f/6$ at the prime or Newtonian focus. Consequently, the collimator of a low-dispersion spectrograph to be used there can be kept short, and the spectrograph as a whole is a fairly small and compact piece of equipment.

A low-dispersion spectrograph, mainly for work on faint nebulae, has been built for the prime focus of the Crossley 36-inch reflector of

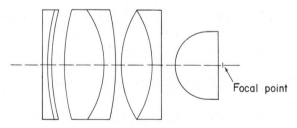

Fig. 141.   Rayton lens.

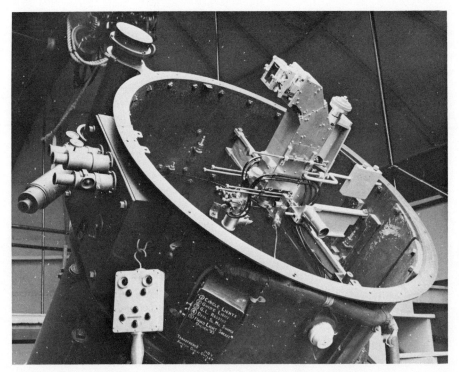

Fig. 142.   Mayall's nebular spectrograph. (Courtesy Lick Observatory.)

the Lick Observatory (Fig. 142). The camera lens has an aperture ratio of $f/1.3$. Objects to be observed with it are sometimes too faint to be seen on the slit by the eye of the observer, and guiding of the telescope has to be done with a nearby field star of adequate brightness. A mechanism is provided to orient the spectrograph properly and for guiding it with a brighter star, whose angular distance from the object to be photographed must be known.

The 200-inch Hale telescope on Palomar Mountain has a low-dispersion spectrograph for its prime focus. Gratings of 300 and 600 lines per millimeter with the blaze for the second-order violet are used alternately. Two cameras of 3-inch aperture are provided, both of the thick-mirror Schmidt type with focal lengths of 1.4 inches and 2.8 inches, and f-numbers of 0.47 and 0.95. The shorter of the two cameras has the smallest focal length permissible for a 3-inch aperture of a thick-mirror Schmidt. The dispersions obtainable with the various grating-camera combinations are 105 A, 210 A, and 430 A per millimeter. This instrument again has provisions for accurately offsetting the slit from a nearby bright star, if the object to be observed is too faint to be seen visually. Radial-velocity work on galaxies between the 19th and 20th magnitudes has been done with this fast spectrograph.

Spectrographs with medium dispersion, in the range of 100 A/mm to 10 A/mm, are usually attached to the Cassegrainian focus (Figs. 143 and 144) if used with a reflector. The f-number is here of the order of $f/15$ to $f/25$, and the collimator becomes rather long, except when it is of the folded type. These instruments are large and fairly heavy pieces. They are fastened to the tube of the telescope.

Both the telescope tube and the spectrograph must assume various positions with respect to the direction of gravity. The spectrograph is consequently subject to flexures that depend on its momentary position. These flexures lead to slight changes in the relative positions of slit, lenses, dispersing element, and photographic plate. Changes of the flexure of the instrument can lead to a shift between the comparison and the stellar spectra, whereby erroneous values of radial velocity would result. Systematic errors thus introduced vary, not only from exposure to exposure, depending on the direction in which the telescope has been pointed, but also during one exposure, if it is of sufficient length.

To prevent such errors, one never exposes the comparison spectrum a single time during the exposure of the stellar spectrum, but rather one breaks the required exposure time of the comparison source into a number of short exposures properly distributed over the longer exposure time on the star. Thus, if the flexure of the spectrograph changes during the exposure time, relative shifts due to differential flexure between the spectra of the two sources can be kept to a minimum by this procedure. Still, a broadening of the lines caused by a slight shift of the spectra on the photographic plate occurs, if the flexure of the instrument has changed from the beginning of the exposure to its end. A rigid mechanical design of the spectrograph is therefore a prerequisite for an instrument supposed to produce sharp lines suitable for wavelength measurements and photometric work.

A stellar spectrograph is subject not only to flexure caused by gravity but also to the influence of temperature. A change in temperature causes the metal body of the spectrograph to expand or to contract, thus changing the distances between the optical components. It also varies the focal lengths of lenses and mirrors, because

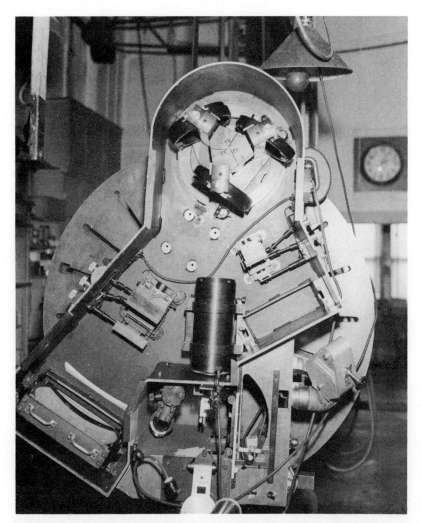

Fig. 143.   Grating spectrograph for the 60-inch reflector of the Mount Wilson Observatory, with folded Cassegrainian collimator (*center*), three gratings, and three Schmidt cameras. (Courtesy Mount Wilson and Palomar Observatories.)

the linear dimensions of these pieces change with temperature. Of much more influence, though, is the change in the dispersing power of the prisms of the spectrograph with temperature. Variations in the dispersion matter only if they occur during the exposure. Even so small a change of temperature as 1° can produce line shifts on the photographic plate that exceed the shifts due to the Doppler effect.

Fig. 144(*a*).   The prism spectrograph at the Cassegrainian focus of the 72-inch reflector of the Dominion Astrophysical Observatory. (Courtesy Dominion Astrophysical Observatory.)

Obviously, no meaningful measurements of radial velocities can be obtained if provisions are not made to maintain constancy of the prism temperature; it should be held within 0.1 C deg during the exposure. Therefore, a jacket around the instrument is usually provided. A thermostat, together with suitably arranged heating elements, establishes a constant temperature of the spectrograph and its components, particularly the prisms, while the photographic

Fig. 144(*b*).   Another view of the prism spectrograph shown in Fig. 144(*a*).

plate is exposed. Temperature changes during the exposure are much less critical in spectrographs equipped with diffraction gratings, and temperature control is not necessarily required.

Operational difficulties encountered with spectrographs due to flexure and temperature changes increase seriously with the size of the instrument and eventually become prohibitive for the construction of large instruments with high dispersions, if these are to be attached to a moving telescope tube.

The most suitable arrangement for a high-dispersion spectrograph is an instrument used at the coudé fixed focus (Fig. 145) on the polar axis of a reflector. The telescope then produces an image in a laboratory room that can be kept at constant temperature. The spectrographic apparatus is essentially stationary. Large optics with long focal lengths can be employed without encountering insurmountable engineering problems. Gratings and other optical components can be exchanged, moved into working position, and properly adjusted by remotely controlled motor drives.

The 200-inch Hale telescope on Palomar Mountain is equipped with a coudé spectrograph (Fig. 146) employing a 12-inch mirror collimator with a focal length of 30 feet. Gratings of a corresponding size cannot be made easily. Recourse, therefore, has been taken to a composite made up of four gratings, each having a ruled area of 5½ by 7 inches.

There are five cameras with apertures of 12 inches for this spectrograph; their focal lengths are 144, 72, 36, 18, and 8.4 inches, and their f-numbers are $f/12, f/6, f/3, f/1.5$, and $f/0.7$. They yield dispersions ranging from 2.3 A/mm to 39 A/mm in the violet region of the spectrum and in the third order. Even with such a powerful instrument as the 200-inch, the limiting magnitude for the highest dispersion is only magnitude 7.5 for an 8- to 10-hour exposure. With the lowest dispersion, stars of magnitude 16.6 can be reached by an exposure of the same length.

A coudé spectrograph planned for the new 120-inch reflector of the Lick Observatory will be equipped with cameras of up to 160-inch focal length. The highest dispersion obtainable will be 0.85

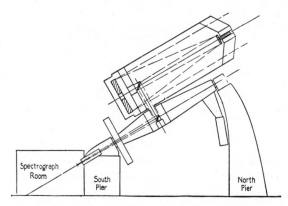

Fig. 145.   Diagram showing the location of a coudé spectrograph.

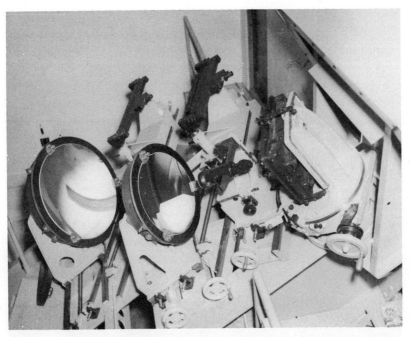

Fig. 146.   Portion of the coudé spectrograph of the 200-inch reflector. The grating holder is to the right. (Courtesy Mount Wilson and Palomar Observatories.)

A/mm in the blue region and in the third order. This high dispersion can be used only for stars brighter than the fifth or sixth magnitude.

For some types of work, efforts are being made to replace the photographic plate in the spectrograph camera by a photocell that can be moved in the direction of dispersion through the spectrum imaged by the instrument. A record is obtained by a chart recorder in the form of a tracing. The introduction of image tubes leads to better and faster recording of stellar spectra without having recourse to the interposition of the photographic process. This technique of recording becomes particularly important for spectrophotometric work. At present, however, the photographic technique is still more efficient than scanning.

Measurement of Stellar Magnetic Fields

In 1896, Peter Zeeman discovered that the spectral lines of sodium were broadened when the source is placed in a strong magnetic field. Shortly after the discovery had been made, Lorentz developed a theory which predicted that the spectral lines should be split into distinct components when produced in a magnetic field. This was confirmed by observational evidence obtained with spectrographs of sufficiently high resolving power. Two line components appear when the source is observed parallel to the lines of force of the magnetic field and three when it is viewed perpendicular to the magnetic field. The light of line components is circularly polarized when seen in the longitudinal direction and plane polarized when viewed in the transverse direction. This is the split pattern of the normal Zeeman effect. Lines are sometimes capable of displaying a much more complicated split pattern (Fig. 147). Occasionally, ten and more components appear. The phenomenon of the anomalous Zeeman effect, as it is called, can be well understood with the help of quantum theory of atomic structure and radiation. The polariza-

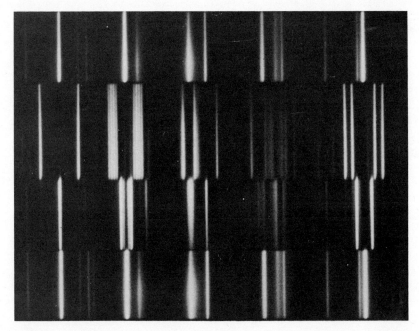

Fig. 147.   Zeeman effect for chromium. At top and bottom is the spectrum from 4612 to 4627 A without a magnetic field. The second strip shows the same spectrum taken under the influence of a magnetic field of 31,700 gauss at right angles to the line of sight, showing only components that are polarized at right angles to the magnetic field. The third strip shows only components polarized in the plane parallel to the field.

tion of the light of the various components is similar to the normal Zeeman effect. The separation of the split components depends on the field strength at the location of the emitting or absorbing atoms.

If a star is surrounded by a magnetic field, we cannot expect that its spectral lines will be split into distinct components. The direction and strength of the magnetic field would vary over the surface of the sphere, and the magnetic effect on the absorbing and emitting atoms would depend on their location on the surface. Consequently, the Zeeman pattern would vary over the stellar surface. Since we observe light integrated over the total surface of the star, we deal with a superposition of many different patterns. No distinct split lines will appear, but the single lines will be somewhat broadened. The light in the two wings of such a broadened line will be circularly polarized in opposite directions. By a suitable technique, the polarization can be utilized for the determination of the magnetic field strength.

H. W. Babcock has successfully employed a device in combination with the high-dispersion coudé spectrograph (2.9 A/mm) of the 100-inch reflector of the Mount Wilson Observatory. He uses two slightly different analyzers in front of the spectrograph slit. One of them consists of a suitably oriented quarter-wave plate of mica with a thickness chosen to be correct for 4600 A, followed by a plane-parallel crystal of calcite. The calcite splits the incident beam into two beams. They emerge parallel from the crystal and are separated by a distance proportional to the thickness of the crystal. Both transmitted images of the star fall on the slit. Because of their separation, two closely adjacent spectra are produced on the photographic plate.

If any right-handed circularly polarized light is received by the analyzer, it will be directed into only one of the two spectra, while left-handed circularly polarized light will be directed into the other. Unpolarized light will be equally divided between the two.

The second analyzer used by Babcock differs from the first in that the calcite crystal is replaced by two pieces of Polaroid with their axes perpendicular to each other and mounted with the dividing line perpendicular to the slit of the spectrograph. As before, a mica quarter-wave plate is used as an analyzer. This latter arrangement again produces two contiguous spectra having the desired properties, if the star images are trailed along the slit, but it is more wasteful of light than the first analyzer.

Since light of different polarization happens to be mainly in the wings of the spectral lines, the separation of the two differently polarized components will lead to a slight shift of the line positions in the two adjacent spectra (Fig. 148). The shift depends on the field strength and is the quantity to be measured. Even on high-dispersion spectra it is of a rather small amount. A field strength of 500 gauss will produce a differential displacement of about 0.001 mm on the plate.

To date, magnetic fields of a number of stars have been found. Some stars display very interesting variable magnetic properties and often great field strengths.

Instruments for Wavelength Measurements

An important part of the evaluation of information contained on a photographic stellar spectrum is the measurement of the positions of the lines. From these positions the wavelengths of the spectral lines can be determined. Identification of the chemical elements

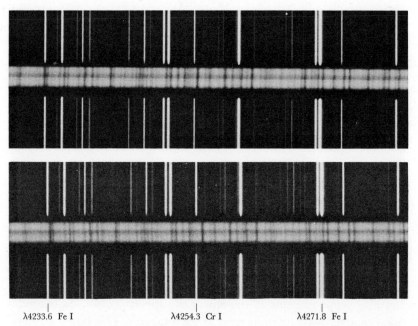

λ4233.6 Fe I   λ4254.3 Cr I   λ4271.8 Fe I

Fig. 148.   Differential line displacements in the spectrum of HD 71866 caused by changes in its magnetic field. Original plates have a dispersion of 4.5 A/mm. Period of magnetic variation of the star is 6.8 days. The effective field intensities were +1940 gauss (*upper spectrum*) and −1870 gauss (*lower spectrum*). (Courtesy Mount Wilson and Palomar Observatories.)

that cause the lines and the measurement of radial velocities with
the help of the Doppler effect are two of the uses to which these
wavelengths can be put.

These measurements are frequently made by means of a measuring comparator. The spectrum is viewed through a low-power
microscope with a reticle in its field. The photographic plate with
the spectrum aligned in the direction of motion can be moved by
means of a precision screw with a divided head. After the spectral
line to be measured has been properly brought under the lines of
the reticle, the screw setting is read. By comparing the readings for
the lines with unknown wavelength with the readings for those
with known wavelength, the unknown can be found.

Such a measuring comparator is shown in Fig. 149. The graduated dial, rotating with the precision screw, permits the reading
of 1/1000 of a revolution, corresponding to a fraction of a micron.
The number of full revolutions of the screw is indicated by the
scale on the front of the instrument. The precision screw is the
heart of the comparator. Its accuracy largely determines the ac-

Fig. 149.   Measuring comparator for spectra. (Courtesy P. M. McPherson Co.,
Acton, Mass.)

curacy of the measurements; its progressive and periodic errors should be negligible. If they exceed the permissible size, corrections have to be applied to the measurements. Also, the bearings for mounting the screw are of the highest importance for the accuracy of performance; no lost motion can be tolerated. The top stage carrying the photographic plate is equipped with a tangent screw to rotate the plate in order to align the spectrum with the direction of motion produced by the precision screw. The longitudinal stage moves and is accurately guided on vee and flat ways. They should be flat and straight to 0.002 mm throughout their entire length. To maintain the utmost accuracy the instrument offers, the temperature of the measuring room should not change during the measurements.

The measurement of a spectrogram for the purpose of determining the radial velocity of a star is a tedious and time-consuming undertaking. Several hundred settings of the micrometer and readings of the dials may be required, in addition to considerable computational work afterward. Radial-velocity work often involves the evaluation of large numbers of spectrograms. Techniques that expedite this work are highly desirable.

One of the first spectrocomparators designed to reduce the amount of work required to derive the radial velocity of a star was constructed by Hartmann. The instrument compares a stellar spectrum suitable for the determination of a radial velocity with the spectrum of a star with well-known velocity. The two spectra appear simultaneously side by side in the eyepiece of the microscope. They can be shifted with respect to each other in the direction of dispersion. First, the two comparison spectra accompanying each stellar spectrum are made to coincide. Then, one of the two plates mounted on their carriages is moved by a screw until the lines of the two stellar spectra coincide. The amount by which one of the two spectrograms has to be moved to make the lines of both match, after the comparison lines of the artificial source have been brought to coincidence, is a measure of the difference in radial velocity of the two stars. Since all that has to be measured in this method is a single small quantity, measurements can be made quickly, and the mechanical requirements for the screw of the instrument are not very stringent, though periodic errors of the screw are undesirable.

Various modifications of these two types of instrument exist. To

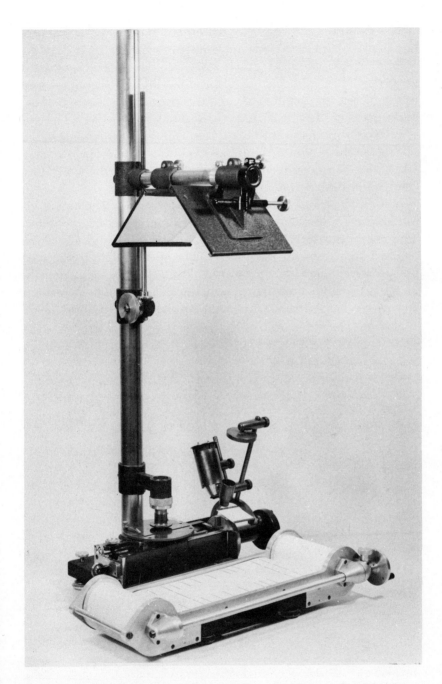

Fig. 150. Spectral comparator of the projection type. (Courtesy Dominion Astrophysical Observatory.)

reduce eyestrain and fatigue of the observer, devices using projection have been developed. Such instruments are in use at the Dominion Astrophysical Observatory in Victoria, B. C., where a great deal of radial-velocity work is done. In a projection version of the measuring comparator, the spectrum and an image of the micrometer head with its index is focused on a screen. The observer performs the measuring process, including the reading of the setting, on this screen without looking into a microscope eyepiece. Otherwise, the instrument corresponds closely to the measuring comparator described above.

A somewhat more advanced type of projection instrument for radial-velocity work resembles the Hartmann spectrocomparator. This machine avoids the use of visual microscopes entirely (Figs. 150 and 151). A highly magnified image of the system is projected upon a screen where it is measured. The photographic plate with the spectrum is mounted upon a turntable of a conventional micrometer with a high-precision screw and a divided micrometer head.

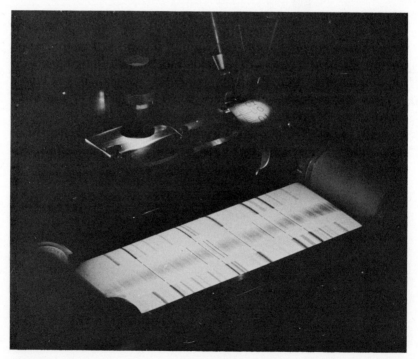

Fig. 151.   Screen of a projection type spectral comparator. (Courtesy Dominion Astrophysical Observatory.)

The viewing screen actually is a scale upon which are ruled the positions of the lines of the comparison spectrum and also those of the stellar lines corresponding to radial velocity zero. The projected image is matched in size precisely to that of the ruled scale and the projected lines are set exactly upon the ruled comparison lines of the scale. This done, the velocity displacements of the stellar lines, with respect to their ruled zero positions, are seen at once and may be measured directly with the micrometer. The amount of displacement is then converted into radial velocity by a simple multiplication by a scale factor. This very direct method involves some danger of personal bias entering the results, since the operator is aware almost immediately of the result while still in the process of measuring. A deliberate effort will be required to secure impersonal settings on the different spectral lines.

For the measurement of high-dispersion laboratory spectra of terrestrial sources, fully automatic computing and recording comparators have been developed. They are not very well suited for the kind of wavelength work to be done in astrophysical spectroscopy and have thus found little use. Yet is seems entirely feasible that such fully automatic measuring machines could be devised for the measurement of stellar spectra.

Recording Densitometers

The photograph of a spectrum contains valuable information about the intensity of the light that the star emits at different wavelengths. The general background of the spectrum, if such exists, is called the continuum. Upon it are superimposed more or less well-defined absorption and emission lines and bands. The intensity distribution through the spectrum enables the astronomer to reach conclusions on the temperature of the stellar atmosphere, its pressure, its chemical composition, and many other quantities. To draw these conclusions by a detailed analysis, the intensity distribution in the spectrum in the direction of dispersion needs to be measured.

The intensity of the stellar radiation produces densities of varying amounts over the spectrogram while the photographic plate is exposed. The way the density varies in the direction of dispersion does not immediately permit a conclusion as to how the intensity of the radiation cast upon the plate varies with wavelength. We have seen that, even for light of the same color, the intensity of the light and the photographic density produced by it are not necessarily proportional to each other. Furthermore, the photographic response depends also upon the wavelength or the color of the light, since the sensitivity of photographic emulsions is not uniform for all colors. Many emulsions, for example, are not sensitive to red light at all or only very little. The relation between density and intensity—the characteristic curve—ought to be known for different colors, if densities are transformed into intensities of the light that produced them.

We are not so much concerned here with how these relations can be set up as with the task of measuring how the density along a spectrum varies. In principle, many of the well-known instruments for measuring the brightness of stars on photographic plates can be used with slight modifications to measure the density variation in the direction of dispersion. These instruments do not produce a continuous record over the length of the spectrum, but a sequence of values at discrete points more or less widely separated. These instruments are not able to reproduce all the photometric information contained in a spectrogram. The ideal aim of obtaining this information as completely as possible can be approximated, however, by spacing the points of the spectrum to be measured as closely as possible. The more complete the knowledge of density variation desired, the more tedious and time-consuming becomes the work of measuring.

Consequently, densitometers have been developed that scan the spectrogram and produce a continuous record of the density along the length of the spectrum (Fig. 152). A large variety of instruments have become available which are mostly of the absolute measuring type (p. 162). A beam of light is sent through a well-defined area of the photographic plate and the reduction of its intensity is determined. Ordinarily, the restricted area is small, and the instrument is often referred to as a microdensitometer.

The main feature that distinguishes densitometers for spectral

Fig. 152.   Direct-intensity tracing of a high-dispersion stellar spectrum.

work from those used on stellar images is a mechanism which moves the plate upon a carrier across a scanning slit of small width by an accurate screw. The scanning beam passes through this slit and the photographic plate. The output of the photocell or any other light detector is recorded photographically on sensitive paper or film, or with the help of a recording meter on a strip chart. The speed of the scanned plate and the recorder chart must be kept carefully at a constant ratio to maintain a defined ratio between the dispersion of the spectrogram on the plate and the dispersion of the tracing on the chart. This ratio can usually be changed by changing the gear reduction, or by other means. By a suitable choice of the magnification, fine details on the plate can be enlarged to the desirable length to permit detailed study. If a high scanning speed is used, caution must be exercised to assure that the recording device is still capable of following rapid density changes without distorting or smoothing out any detail. The resolving power of the instrument should be higher than that of the photographic plate in order to make the finest details of the photographic plate visible on the tracing. This requires a narrow scanning beam and consequently high sensitivity of the light detector of the densitometer.

Many of the older densitometers are built on a basic design originally used in instruments proposed by Koch and by Moll. In Koch's recording instrument a photocell and an electrometer serve to measure and record the density of the plate. Moll's instrument uses a thermocouple and a galvanometer. In more recent constructions, these detectors are replaced by photomultipliers.

The accuracy of the tracing depends very much on the accuracy with which the motion of the plate and the chart can be sychronized. Gear trains may easily introduce inaccuracies by lost motion or gear errors; hence they are frequently replaced in the transport mechanism by lever arrangements. More recently, two synchronous motors depending upon the same alternating current have been employed. One of them drives the plate under the scanning beam, and the other moves the strip chart. Frequency fluctuations of the current do not affect the speed ratio of the two motions, thus assuring a constant relation between the dispersion of the spectrum and that of the tracing.

The Baird recording densitometer (Fig. 153) utilizes a two-beam optical null method; the tracings thus are unaffected by brightness fluctuations of the light source. The light source (Fig. 154) feeds

Fig. 153.   Baird densitometer, recorder not shown. (Courtesy Baird Atomic Instruments Co.)

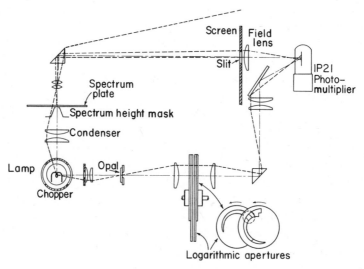

Fig. 154.   Optical schematic of Baird densitometer. (Courtesy Baird Atomic Instruments Co.)

light into the sample and measuring beam. A slit transmits the light that has passed through the sample spectrum to a photomultiplier. An aperature in front of the multiplier excludes from the photocell all light except that which is transmitted by the projector.

An electronic system operates as a self-balancing null indicator in which the intensity of light in the reference beam is automatically and continuously adjusted to make it equal the intensity in the measuring beam. A chopper alternates the two beams to the photomultiplier. The light falling on the photocathode is steady, with no flicker, when the intensities of the two beams are equal. When the two intensities are not equal owing to changed density of the plate where the measuring beam falls, the flicker incident on the multiplier cathode produces an alternating-current signal output. This signal, after amplification, is supplied to a balancing servomotor. This motor drives the precision logarithmic-aperature shafts until the transmission of the reference-beam apertures again matches that of the sample plate in the measuring beam. The density scale is geared to the servomotor that drives the apertures. By a slight modification the position of the shaft of the density scale can be transmitted to a recording meter that puts tracings on a chart.

After the scanned record of the spectrum has been obtained, it is usually necessary to transform this curve, representing photographic density as a function of wavelength, into a curve representing intensity of incident light as a function of wavelength. The deflections represented by the curve have to be read at many closely spaced points. With the help of the characteristic curve of the plate, valid for the particular wavelength region, the readings must be transformed into intensities and then plotted again. If long, complex tracings, or many tracings, are to be evaluated, the labor involved in the reduction may be tremendous, or even prohibitive for certain problems. Various attempts have been made to reduce the drudgery of this procedure. Two quite different approaches suggest themselves.

The first one uses the density tracings as produced by the instrument. A more or less automatic auxiliary instrument rapidly accomplishes the graphical reduction into an intensity record. Information on the characteristic curve of the plate must be fed into the device. Speed, as well as ease of operation, is an essential requirement. The accuracy of which the device is capable should be high enough not to distort the information contained in the photographic emulsion.

The other solution utilizes recording instruments which do not record photographic density at all, but which immediately produce a record in terms of intensity with the help of the characteristic curve.

A semiautomatic instrument for the graphical transformation of density into intensity curves has been described by Beals. Some of the same principles have been utilized in instruments suggested by others.

A calibration curve P (Fig. 155), showing deflection on the strip chart as ordinate versus intensity as abscissa, is drawn on a ground-glass plate G, illuminated from below. This glass screen is mounted on a carriage which can be moved freely to the right and left by the operator. A wire under the plate casts a shadow on the screen; this shadow remains stationary and acts as an index I. A lens and

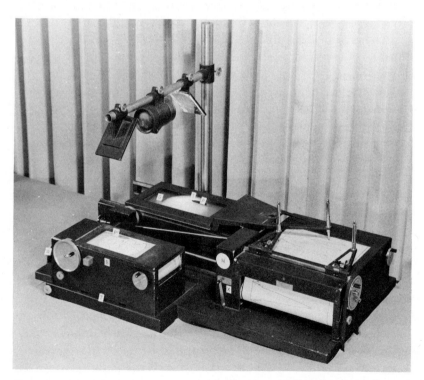

Fig. 155.   Semiautomatic density-intensity converter. (Courtesy Dominion Astrophysical Observatory.)

two 45° mirrors project the calibration curve and the index upon a microdensitometer tracing P' slowly moving from right to left. The operator continuously adjusts the position of the movable carriage so that the point of intersection of the calibration curve and the vertical index always falls on the tracing.

Attached to the moving carriage is a pen that bears upon a paper strip moving synchronously with the tracing. To every deflection on the tracing, indicated by the intersection of the vertical index with the tracing, there corresponds an intensity (abscissa) indicated by the position of the pen on the moving paper strip. The pen draws continuously the intensity curve while the operator keeps the calibration curve and the tracing intersecting the index at the same point.

The second solution, a densitometer that records, as a function of wavelength, the intensity of a source instead of the photographic density of the plate, has been realized in various instruments. No manual or automatic conversion by an intermediate procedure is required; the recording instrument itself accomplishes the conversion, and the tracings immediately represent the intensity variation.

M. Minnaert and J. Houtgast solved the problem by an optical arrangement added to a Moll microdensitometer. The device was first used to work out a photometric atlas of the solar spectrum at the Utrecht Observatory. The atlas gives a complete intensity record of the spectrum of the sun between 3600 A and 8800 A wavelength as obtained from high-dispersion spectra. Without a direct-intensity recording device the tremendous task of making this atlas never could have been accomplished.

In the Utrecht instrument the output current of the light detector is conducted to a galvanometer G_1 (Fig. 156), whose mirror is illuminated by light from a source L_1 that passes through a long slit S. After reflection by the mirror, the light passes through a diaphragm D cut in the shape of the calibration curve of the particular spectrum under investigation. An image of the slit is formed on this diaphragm. For every density of the photographic plate the galvanometer mirror is deflected through an angle proportional to the plate transmission. The diaphragm passes a part of the luminous slit image that is proportional to the intensity. All light transmitted by the diaphragm is focused by a lens A on a photocell (the authors used a blocking-layer cell). The output of this cell is led to the galvanometer G_2 of the densitometer, and its deflection is photo-

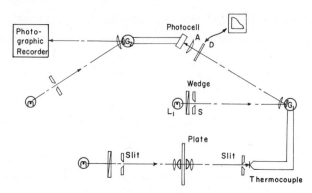

Fig. 156.   Scheme of a direct-intensity densitometer used at the Utrecht Observatory.

graphically recorded. The records could, of course, be made by an ink recorder.

A photographic wedge of variable transmission is inserted in the beam between the light source L_1 and the slit in order to adjust the light level of the source so that the galvanometer deflections do not exceed the chart width of the recorder. Before the tracing work can be done, the characteristic calibration curve has to be determined and the diaphragm cut out.

A direct intensity-recording instrument based on rather different principles has been developed by R. C. Williams and W. A. Hiltner. It produces tracings on either an intensity scale or a logarithmic intensity scale. The device has two separate plate carriages, both of which can be driven simultaneously along parallel ways at equal or varying rates of travel. One carriage, called the balance carriage, can also be moved in a transverse direction. On one carriage is placed the plate with the stellar spectrum to be analyzed. The other one carries a calibration spectrum across which the intensity is known to vary exponentially with distance. Slit images of two light sources are focused on the two plates at the same wavelength; both spectra are taken with the same spectrograph.

By suitable optical, electrical, and mechanical means the balance carriage is moved transversely until the density of that part of the stellar spectrum under investigation is equal to the density of the calibration spectrum. The motion of the balancing carriage is transformed by means of a cam and a mirror to the motion of a recording light beam across a drum containing photographic recording paper. The cam is so shaped that the distance moved by the spot of light across the recording drum is proportional to the relative intensity of the light that originally impressed itself upon the calibration spectrum. Since the calibration spectrum and the stellar spectrum are kept together in both wavelength and density, the motion of the spot of light on the drum is proportional to the relative intensity of light which impressed itself upon the stellar spectrum, and direct intensity tracings result. The effect of the calibration spectrogram is always introduced at the right wavelength, since the two carriages are driven so as to keep together in wavelength. An alternate cam is also provided in case tracings with a logarithmic intensity scale are wanted.

7

Instruments for
Solar Research

Study of the sun is of special importance to astrophysics because the sun is a rather typical star and it is so close to us that it can be subjected to detailed scrutiny. As we have seen, other stars are so far away that we can hardly hope to study the structure of their surfaces. On the other hand, spots on the nearby sun are sometimes so huge that they can be seen by the eye aided only by dark glass. With special instrumentation an infinite wealth of detail becomes discernible. On the rare occasions of total eclipses the awe-inspiring and beautiful solar corona and prominences are visible. Many special instruments have been developed for viewing and studying these phenomena without the benefit of an eclipse.

Because of the unusual equipment required, a number of observatories are devoted exclusively to the study of solar phenomena. In the United States the sun is the principal concern of the McMath-Hulburt, the Sacramento Peak, and the High Altitude Observatories, and constitutes a large part of the work of the Mount Wilson and Palomar Observatories. In the rest of the world, much solar research is done in Australia, Austria, Belgium, Chile, Czechoslovakia, England, France, Germany, Greece, Holland, India, Ireland, Italy, Japan, Peru, Scotland, Spain, Switzerland, Turkey, and the U.S.S.R. The sun is, therefore, under nearly continuous observation. In recent years relations between solar changes and terrestrial phenomena such as magnetic disturbances, radio fade-outs, and cosmic-ray bursts have been definitely established.

Direct visual observation of the sun can be performed with many telescopes after protective devices have been added. One of the simplest ways that the sun can be observed with a visual refractor is by projection. A white card is placed about 10 inches behind the eyepiece (depending on the size of solar image desired) and the eyepiece is adjusted to focus the disk on the card. In this way drawings of the disk may readily be made, and this projection method is perhaps the best way of recording the often numerous small sunspots. Photographs may also be taken by replacing the card with film and providing a lighttight box and a shutter.

Another way in which the disk may be observed is by using an eyepiece in the normal manner but with the addition of a Herschel wedge (Fig. 157) and solar filter. Unless the telescope aperture is very small, a filter should not be used alone, even if sufficiently dark. The intense heat near the focus may cause the filter to break

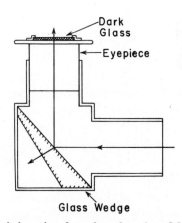

Fig. 157.   The Herschel wedge for solar observing. Most of the light passes through the wedge. The part that is reflected from the front surface forms the image that is observed. The part that is reflected from the back surface is stopped within the housing.

suddenly and admit the full sunlight to the eye. Furthermore, though glass filters may appear to be quite dark in visible light, they may not be so dark in the infrared. Though the eye may seem comfortable while viewing the sun, the retina is being cooked and may be irreparably damaged by the heat. To prevent these misfortunes the wedge should be used. As the figure shows, the sunlight is reflected from the front surface of the glass. The back surface is inclined at another angle so that the reflection from this surface will be thrown to the side and will not interfere. An absorbing glass filter must be used along with the wedge, as the light reduction of the latter is not sufficient for viewing in comfort.

In general, atmospheric turbulence is greater in the daytime than at night as a result of solar heating. Thus day seeing is almost never as good as night seeing. One might think, too, that the best solar seeing would be obtained at noon, because the atmospheric path is then at a minimum. But in reality it is found to be poor during this part of the day because of atmospheric turbulence due to heating of the ground. The best observing conditions are generally found not long after sunrise, when the sun has attained sufficient altitude that refraction does not distort it appreciably. At about this point the atmospheric path is not unreasonably long, and usually the turbulence has not increased unduly. Observations of the solar disk and particularly the recording of sunspots are best performed at this time of the day.

Tower Telescopes

It was discovered by George E. Hale that during the day the seeing at some distance above the ground is generally better than near to the surface. He found, too, that an underlying tree cover is beneficial. In order to take advantage of these circumstances and to provide a telescope to feed a very high-dispersion spectrograph, he devised the tower telescope (Fig. 158). At the top of a 150-foot tower (Fig. 159) there is an equatorially mounted flat mirror which, usually with a secondary flat mirror, provides a stationary beam of parallel sunlight which travels vertically down the inside of the tower to a laboratory beneath. The radiation is thereby collected above the level of the poorest seeing and travels down an enclosed, thermally insulated pipe, where only small turbulence is present. A lens or mirror objective of long focal length is often

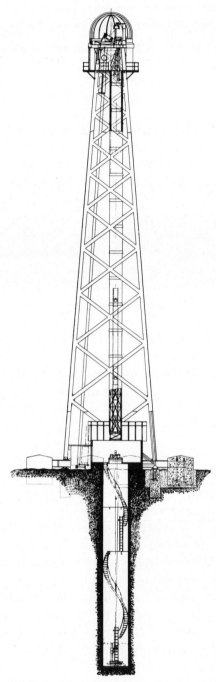

Fig. 158.   Drawing of the 150-foot tower telescope and its spectrograph. (Mount Wilson and Palomar Observatories.)

Fig. 159.   The 150-foot tower telescope of the Mount Wilson Observatory. (Mount Wilson and Palomar Observatories photograph.)

placed at the top of the tube to provide an image in the laboratory beneath. The mirrors and the objective are housed within a dome. The slit is opened and the mirrors are initially adjusted from the top of the tower. There are controls in the laboratory beneath for rotation of the dome and fine motion of the solar image.

There are three different arrangements of mirrors and driving motions which provide a stationary solar image. They are known by the names siderostat, heliostat, and coelostat. Only the last will be described here, as it is the one most often employed.

The coelostat (Fig. 160) has a flat mirror mounted in a cradle so that the polar axis is parallel to and through the center of the face of the mirror. The clock drive rotates the mirror about this axis at the rate of once in 48 hours, that is, at one-half normal solar telescope speed. As is well known, a ray reflected from a mirror moves through twice the angle that the mirror turns through. Therefore, if a celestial object is viewed in the coelostat mirror, it will appear stationary. A secondary mirror is normally arranged to divert a parallel beam of light from the desired object to wherever the light is desired. The declination is changed by altering the relative positions of the primary and secondary mirrors. Since in a tower telescope the position of the secondary mirror is fixed by the objective lens beneath it, the primary mirror together with its drive

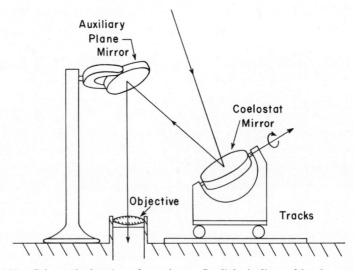

Fig. 160.   Schematic drawing of a coelostat. Sunlight is directed by the two mirrors into the fixed telescope objective.

mechanism is moved on rails along a north-south line to change the observed declination. In addition to this motion, the tilt of the secondary mirror has to be adjusted for each new declination. Furthermore, the tracks for the coelostat mirror must be either east or west of the secondary mirror. Usually both sets of rails are provided, since the secondary mirror structure will shadow the primary at some time during the day if the primary can be used only on one side of the secondary.

The coelostat has the advantage over the heliostat that not only is the center of the image held stationary but the image does not rotate. The siderostat has an advantage over both of the others in that only one mirror is needed, but its driving mechanism, which requires levers, is much more complex.

The mirrors are preferably made from fused quartz to avoid expansion from the solar heating. In addition, if the ultraviolet is not desired, the surfaces should be silvered rather than aluminized as silver has the better reflectivity in the visible and infrared and consequently only about half of the absorption of heat.

High-Dispersion Spectrographs

Probably the most important physical aspect of our nearest star is its spectrum. Because it can be resolved so well and because it can be studied in detail over the surface of the sun, an understanding of the physical conditions in the sun's outer layer is being derived from it. Questions of turbulence, thermal structure, electron density, pressure, atomic scattering, and continuous as well as line absorption are presently being settled. One of the triumphs of solar astronomy was the precise determination of the chemical composition of the sun. Indeed, lines of 67 elements have been seen in the spectrum. But still there is much work to be done. For example, there are about 26,000 lines in the readily photographed part of the spectrum. About 30 percent of these, or 8,000 lines, yet remain to be identified as to their origin. Further, there are shifts in the positions of lines in sunspot areas and near to the limbs of the disk which are not yet well understood.

The dispersion of a stellar spectrograph is usually limited by the small amount of light available. If the dispersion is made too high an exposure cannot be obtained within a reasonable time. Spectrographs for work upon the disk of the sun do not have this limitation, at least not severely. Dispersions as large as 10 mm/A are often used and the dispersion is usually at least as high as 1 mm/A. Spectrographs of this order of dispersion (and resolution, for there is little point to high dispersion without accompanying resolving power) require special considerations in their design.

The *f*-number of a tower telescope may be *f*/100 or greater to provide a large image of the sun. The collimator of the spectrograph, therefore, must have an equally large *f*-number to match the telescope. Since high linear dispersion is desired, a large camera *f*-number is needed as well and the collimator objective often serves also as the camera lens. Diffraction gratings are almost universally used because of the large dispersion they afford. The grating is placed behind the collimator-camera lens in the arrangement known as the Littrow grating spectrograph (Fig. 161), and the plate or exit slit is in the same plane as the entrance slit. Such a spectrograph is often placed vertically in a well beneath a tower telescope. The vertical arrangement offers the advantage over a horizontal one of having fewer effects on seeing. The long focal lengths required for the high dispersion make the atmospheric turbulence occurring within a laboratory a severe detriment to high resolution. The spectrograph is housed within a deep well to insure constancy of temperature and thus to reduce turbulence of the air. Usually the slit is near ground level and so the depth of the well is a little more than the focal length of the collimator-camera lens. At the 150-foot tower the longest focal length is 75 feet.

One way in which the turbulence problem can be eliminated is by removing its source—the atmosphere within the spectrograph. This has been done in a horizontal spectrograph at the McMath-Hulburt Observatory (Fig. 162). The spectrograph is constructed within an iron pipe of 4-foot diameter and 52-foot length and is evacuated by a pump. The all-mirror optical system has a grating

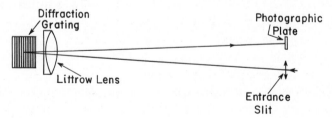

Fig. 161.   The Littrow spectrograph provides very high dispersion and resolution for solar studies.

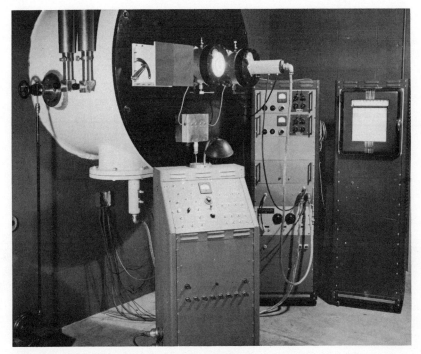

Fig. 162. The slit end of the McMath-Hulburt 52-foot vacuum spectrograph. (Courtesy of *Sky and Telescope*.)

which provides a dispersion of 11 mm/A on the photographic plate. In addition to photography, photoelectric recording of the spectrum can be used. The spectrograph has shown a proved resolution of 600,000.

The Spectroheliograph

At the total solar eclipse of 1868 Janssen made an important discovery for astrophysics and solar instrumentation. The pink-lavender appendages of the sun called prominences had long been seen at solar eclipses but the nature of their light was uncertain. During the eclipse Janssen pointed his spectroscope at the solar limb and found that the prominences had a number of bright lines. He quickly recorded the settings for as many of the lines as he could to obtain their wavelengths. He was much impressed by their brightness, and thought that he might be able to see them in his spectroscope without an eclipse. Indeed, the very next day he tried and found that he could. A few days later Lockyer independently made the same discovery without the benefit of an eclipse.

Though the total light from a prominence is small compared to the immense amount scattered from the solar disk near to the sun, the concentration of light in a few wavelengths permits a prominence to be visible against the background of scattered light if just one of these wavelengths is observed. After Janssen's discovery Huggins showed that simply by opening the entrance slit to encompass an entire prominence its structure can be seen in the eyepiece of the spectroscope. But, unfortunately, the structure of the whole disk of the sun cannot be seen at once by this method. It remained for Hale and Deslandres independently in 1890 to construct equipment for viewing the disk.

The instrument they developed, the spectroheliograph, permits the sun to be photographed in the light of just one wavelength, which may be chosen anywhere but it is usually chosen to fall upon one of the strong lines of the solar spectrum, the H or K lines of calcium or one of the Balmer lines of hydrogen. The spectroheliograph is a spectrograph in which a second slit is placed so that only the light from a desired line, for example the first Balmer line Hα at 6563 A, passes through it. The solar image is arranged to sweep across the entrance slit and at the same time a photographic plate is moved in unison behind the exit slit. Since only light of Hα can get through the spectrograph, an image of the sun in this wavelength is built up on the plate. A photograph of the solar disk taken with a spectroheliograph in the light of hydrogen is shown in Fig. 164. Prominences may be photographed by occulting the sun with a disk before the entrance slit of the spectrograph. This is necessary because of the scattering of the bright light of the solar disk that would occur within the spectrograph.

If the entrance slit is made to sweep rapidly enough across the sun's image and the exit slit is made to sweep similarly across the field of view of the eye, the prominences may be made visible to the eye. The device for doing this is called a spectrohelioscope. The sweeping may be performed (Fig. 163) by oscillating the entrance and exit slits of a Littrow spectrograph in their plane. The visual instrument was invented by Hale in 1924 but is little used today; the polarizing monochromator has been found to be more convenient, being much more compact.

Robert R. McMath at the McMath-Hulbert Observatory intro-

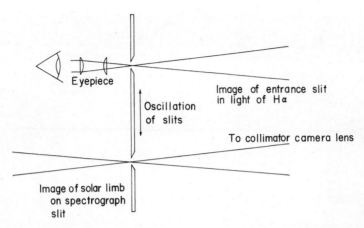

Fig. 163.   The spectrohelioscope is a Littrow spectrograph equipped with oscillating entrance and exit slits and an eyepiece arranged for viewing the image at the exit slit.

duced motion-picture photography to record some of the transient phenomena seen in the spectrohelioscope. Some of the phenomena that can be seen in hydrogen light are plage areas (Fig. 164), changes in the shape and size of loop prominences (Fig. 165), and other changes in the solar atmosphere. The motion of usually slow-moving prominences is emphasized by compressing the time by a factor of about 500. Individual frames are taken at intervals of 15 seconds to 5 minutes, depending on the rapidity of the phenomenon, and projected, after development, at normal speed. Many spectacular changes in prominences are seen if this is done. Much of the motion-picture photography is done today with a polarizing monochromator. Regular patrols are performed at many observatories using such filters to observe flares. These have been found to affect, sometimes profoundly, terrestrial radio transmissions. The occurrence of flares is used to forecast radio propagation.

The Polarizing Monochromator

A simple, compact filter that could be made for even one spectral line, the Hα line for example, would be a most useful piece of equipment. Glass and gelatin filters may be made with bandwidths as small as 50 A, but only at certain wavelengths for which dyes are available. Moreover, the bandwidth is at least ten times too wide to be useful.

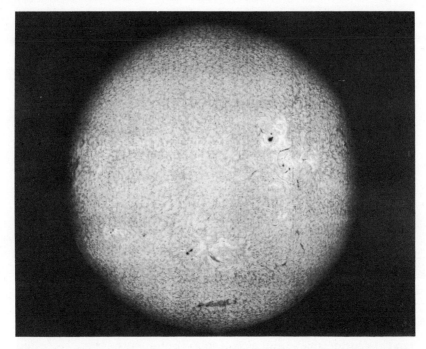

Fig. 164.   Photograph of the solar disk in Hα light with the spectroheliograph, showing plage areas and dark filaments. (Mount Wilson and Palomar Observatories photograph.)

Fig. 165.   A loop prominence photographed with the quartz monochromator. (Sacramento Peak Observatory photograph.)

The polarizing filter was invented by R. W. Wood in 1914, but its present form was suggested by Lyot in 1933 and as a result it is known as a Lyot filter. Öhman built the first practical filter of quartz in 1938. It had a bandwidth of 20 A. A narrower one was made by Lyot in 1944.

Quartz is birefringent, that is, it has two refractive indices, depending on the direction of polarization of the light, except when the light enters it in one direction, called the optic axis. If a plate is cut from a quartz crystal so that a face of the plate is parallel to the optic axis of the crystal, then light entering the plate perpendicular to its face will be split into two beams; in one of them the light is polarized along the direction of the optic axis and travels at a certain velocity through the plate, and in the other the light is polarized at right angles to the axis and travels at a higher speed through the quartz. If a Polaroid is placed before the plate of quartz with its direction of polarization at 45° to the optic axis of the quartz, the intensities of the two oppositely polarized beams of light within the quartz will be the same. After emerging from the quartz plate, the light in the slow beam will have gone through considerably more vibrations than that in the fast beam. As a result of the optical path difference the two beams can be made to interfere with one another (Chapter 1) by bringing their planes of polarization together. This is performed with another Polaroid, again at 45° to the optic axis of the quartz. If the light of a continuous source, after emerging from the second Polaroid, is examined with a spectroscope, the spectrum will exhibit fringes, as diagramed in Fig. 166a. The number of interference fringes per 100 angstroms depends on the thickness of the quartz. For a plate 1 cm thick, there will be about 44 A between bright fringes at Hα, while for a plate twice this thick there will be only 22 A between the fringes.

In order to make a filter to isolate the light of Hα, we must eliminate all of the fringes except the one that falls on Hα. A series of quartz plates (Fig. 167) with other thicknesses will suffice. Suppose we want a monochromator with a 2.2-A bandwidth. First we get a piece of quartz with a thickness of 10 cm and with its optic axis parallel to one face. This will give, between polarizers, fringes at every 4.4 A and, if the plate is accurately ground and polished to the correct thickness, a bright fringe will be exactly on Hα. We add to this a plate of half the thickness and another Polaroid. This plate will have fringes 8.8 A apart and a bright one

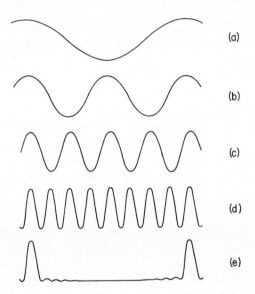

Fig. 166. Transmission of quartz plates between Polaroid sheets for (*a*) the thinnest plate, (*b*) a plate twice the thickness, (*c*) four times the thickness, (*d*) eight times the thickness, and (*e*) all of the plates and Polaroids.

on Hα (Fig. 166*b*). The dark fringes next to Hα will, however, suppress the bright fringes falling at the same place from the first quartz. If we add another slab of one-fourth the thickness of the first, we shall still further emphasize Hα and suppress nearby maxima. Depending on the bandwidth chosen, about six or seven slabs are needed in the relative thicknesses 1, ½, ¼, ⅛, ¹⁄₁₆, ¹⁄₃₂, . . . The result of four plates will be as indicated in Fig. 166*e*. With a six-element filter the nearest bright bands will be 64 times the bandwidth away, about 140 A in the case of the 2.2-A filter. These spurious bands can be eliminated by auxiliary glass or gelatin or

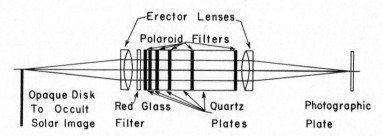

Fig. 167. Arrangement of the plates in the polarizing monochromator.

interference filters. A filter having a band pass of 5 A requires about 9 cm of quartz in six plates and needs seven polarizers.

In addition to quartz, polarizing filters have been made from calcite and ammonium dihydrogen phosphate. These two materials result in thinner filters for the same bandwidth or, alternatively, narrower bandwidths for the same thickness. However, the materials are much harder to work than quartz. The latter has found most application, and many quartz filters are now in use. Construction of quartz polarizing monochromators is a difficult task, but it has been undertaken successfully even by amateurs.

The exact wavelength of the band transmitted by a filter is sensitive to temperature changes. On this account the temperature of a filter must be controlled to within a fraction of a degree centigrade. Alternatively, the filter may be tuned a few angstroms by altering the temperature. However, the response of the filter is slow. The wavelength of the band may also be changed by a slight rotation of the Polaroid next to the thickest element. Doppler shifts, resulting from line-of-sight motions of prominences, can be found in this way.

A polarizing filter is essentially a one-wavelength device, in contrast to the spectroheliograph, which operates at any wavelength. However, quartz filters have been made that will serve several lines of importance to solar physics. One such filter, at Sacramento Peak Observatory, can be used on the 5303-A and the 6374-A iron lines in the corona and the 5876-A helium line, as well as Hα, by altering the auxiliary filter and the operating temperature slightly. The multiline filters result from choosing the thickness of the elements so that some of the other pass bands of the filter fall on or near the wavelengths desired.

Interference filters, as have been mentioned, are a useful adjunct to polarizing monochromators and other instruments of astronomical research. They consist of evaporated layers of magnesium fluoride, zinc sulfide, and sometimes silver. Early interference filters had two thin semitransparent layers of silver separated by an evaporated layer of magnesium fluoride. The thickness of the separating layer was one-half wavelength for the color desired, or some whole-number multiple of this thickness. Part of the light passing through the filter is reflected many times back and forth across the layer separating the semitransparent, semireflecting coatings on the surface. At each reflection a little of the light escapes. In the forward direction escaping light whose wavelength is twice the thickness of the separating layer constructively interferes with each of the previously escaped portions of light, and maximum transmission at this wavelength occurs. At other nearby wavelengths destructive interference occurs and little light is transmitted. The bandwidths of the early filters were a few hundred angstroms and the band could be placed wherever desired. Recent filters have been made using alternate layers of materials of high and low refractive index in place of the silver layers. They provide not only higher light transmission (about 70 percent, twice that of the silver filters) but narrower bandwidths, of the order of 70 A. The exact wavelength transmitted by an interference filter may be changed several hundred angstroms by inclining it to the light beam. The change is always toward shorter wavelengths.

The Coronagraph

Normally the aureole of scattered light about the sun's disk hides one of the most beautiful and scientifically intriguing parts of the sun from our view. But during a total eclipse, when the intense light of the disk is hidden by the moon, the nebulous corona shines out in full glory. Astronomers today are still uncertain as to the origin of the corona, though they have learned much about it. To be able to study this part of the sun without the benefit of rare eclipses is of great importance. The brightness of the corona, however, is only about one-millionth of that of the entire sun. The average sky scatters light near to the sun to the extent of about one thousand times this amount. Fortunately, at some locations, on mountain tops for example, the scattered light is usually below several hundred millionths of the disk and often only twenty millionths. On rare days the aureole of scattered light is only several times the intensity of the corona. With such conditions there is a chance that one might observe the corona without a total eclipse.

An ordinary telescope scatters enough light itself to make coronal observation hopeless. Bernard Lyot succeeded in doing what a number of astronomers had tried but failed to do—observe the corona without an eclipse—and he accomplished this on a fairly routine basis. To achieve success, he moved to the Pic du Midi (2862 m) in the Pyrenees and built a very specialized telescope from which most of the scattering was eliminated. He further utilized the fact that some of the light of the corona is concentrated into spectral lines and he photographed the corona in their light. In addition, he was able on the rare days that the atmospheric aureole was extremely small to photograph the white-light part of the corona.

Lyot's special solar telescope, the coronagraph, is very different from the ordinary one, which tends to obliterate the corona through scattering due to minute scratches and dust on the lens, to bubbles within the glass, to diffraction because of the finite aperture, and to reflections between the surfaces of the lens. Flint glass, needed to achromatize a lens, has many more bubbles than crown and also scatters throughout its volume. Further, in a doublet lens there are reflections between the surfaces. Accordingly, the objective of the coronagraph is only a single lens made from a crown glass chosen for its freedom from bubbles and other imperfections. It is polished with extreme care so that it is free from all scratches, even the most minute.

The sun's disk is artificially eclipsed in the coronagraph by a polished metal cone (Fig. 168). The shiny cone reflects the sun's light to the side so that it cannot fall back on the objective and get into the optical path again. The cone is supported by a rod from the field lens, which serves to image the objective upon a second objective. Here a trick is played on the diffracted light. If one were to look in the coronagraph beside the occulted image of the sun one would see the edge of the objective shining from the light it is diffracting. The trick is to prevent this light from getting into the second objective and it is accomplished by placing a stop at the point where the main lens is imaged by the field lens. A further bit of trickery is to place an opaque disk at the center of the second objective. This disk catches light that is multiply reflected within the primary and also any light that is reflected by the center of the cone to the objective.

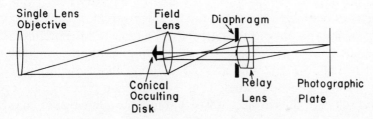

Fig. 168.   Diagram of the Lyot coronagraph.

Chromatic aberration of the primary lens has been neglected. The secondary is overcorrected for color so that the final image is free of aberration, but at the occulting disk or cone the aberration is not corrected. The consequence is that the light from the sun's disk is completely eliminated only for a small range of wavelengths that come to focus near to the position of the cone. Hence the coronagraph must be used with a color filter, but this is a relatively small nuisance. The range of wavelengths may be shifted by moving the occulting disk.

The lens must be kept spotlessly clean. It normally is placed at one end of a dust tube of some length which is coated on the inside with a sticky substance to catch any dust particles. When the instrument is not in use the dust tube is kept tightly capped. With care in keeping the lens clean the instrumental scattering can be kept below the limit set by the atmospheric aureole, even on the best days.

Because of the difficulty in making the main lens, coronagraphs are not large instruments. The largest (Fig. 169) is at the Sacramento Peak Observatory, altitude 2760 m, and has a 16-inch aperture. A similar one is installed at the High Altitude Observatory at an altitude of 3500 m.

The chief application of the coronagraph is to spectroscopy of the corona. Figure 170 shows a spectrum of the corona and also some prominence lines. The corona has strong emission lines in the green at 5303 A and in the red at 6374 A and 6702 A. From a study of the spectrum, much has been learned about the temperature and electron density in the corona.

Photographs of the white part of the corona have seldom been taken except at an eclipse. Only on the rare days when the sky scattering is at its lowest and the corona is very bright was Lyot able to detect the corona photographically. Photographs are often made in the light of one of the emission lines. One of the wavelengths used is the green 5303-A line of the iron atom with 13 electrons removed. The green line is much brighter than the white corona and can be detected fairly easily, often when the brightness of the sun's aureole is as much as 200 millionths that of the disk. A quartz monochromator isolates the radiation and photographs of the corona may be readily made (Fig. 171). Motion pictures have even been taken of the corona and they sometimes show rapid changes taking place.

Fig. 169. The 16-inch coronagraph of the Sacramento Peak Observatory. (Sacramento Peak Observatory photograph.)

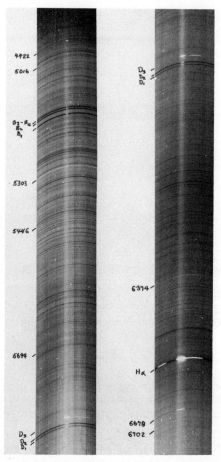

Fig. 170.   The spectrum of the corona and prominences showing numerous emission lines. The curved lines result from the circular slit used to conform to the solar limb. (Sacramento Peak Observatory photograph.)

The Solar Magnetograph

The magnetic fields surrounding sunspots have long been measured in a way similar to that described in Chapter 6 for stars. It is found that, near sunspots, fields between 1000 and 3000 gauss are experienced.

It had long been suspected that the sun had a general magnetic field as well. Photographs of the corona at an eclipse often show streamers or rays which emanate from the north and south poles of the sun and suggest pictures of the magnetic lines of force that emanate from a magnet's poles. Until quite recently the most sensitive tests for a general field were all negative or gave conflicting results. However, in 1952 equipment was developed by Horace W. and Harold D. Babcock that enabled them to accurately map the sun's general field (Fig. 172).

They used the high-dispersion spectrograph at the Hale Solar Laboratory in Pasadena, California and later, after the equipment was well developed, the spectrograph at the 150-foot tower at Mount Wilson. In principle, they detect the field in the same way that stellar fields have been measured. When light of the wavelengths selected for use in this study is emitted in a strong magnetic field, such as that overlying a sunspot, the spectral lines are split into three components whose separation depends on the field strength. The Babcocks have used the line of iron at 5250.2 A. If the field direction is exactly along the line of sight, only the two outer components will be found; the central one disappears. Circular polarizers verify that the two remaining components have right- and left-hand circular polarization. The component that has the right-hand polarization depends on the polarity of the magnetic field, and so the sense of the field can be determined. If

Fig. 171.   Photograph of the corona in the light of the 5303-A line of iron. (Sacramento Peak Observatory photograph.)

Fig. 172. The magnetograph of the Hale Solar Laboratory. (Courtesy of Dr. Horace W. Babcock.)

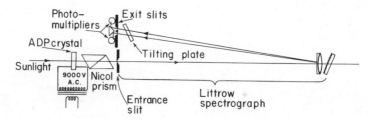

Fig. 173. Schematic diagram of the Babcock solar magnetograph.

now the right-hand polarizer is alternately interchanged with a left-hand one, the line will appear to shift back and forth. The shift is quite easy to see in the vicinity of sunspots, whose field may be 2000 gauss. On the other hand, the shift when the magnetic field is only 1 gauss is but 0.00008 A, while the width of the line is 0.1 A. The Babcocks constructed equipment to measure photoelectrically even smaller shifts than that due to 1 gauss.

Figure 173 shows a diagram of their equipment. The sunlight first traverses a plate of ammonium dihydrogen phosphate (or ADP), then passes through a Nicol prism to the slit of a Littrow spectrograph having a resolution of 600,000. At the focal plane the light from the two sides of the magnetically sensitive 5250.2-A line falls on two photomultipliers. Figure 174 shows that each of the tubes receives light from one wing of the line. A potential difference of 9000 volts applied to the ADP plate makes it sufficiently birefringent to become a quarter-wave plate. The plate and the Nicol prism together constitute either a right-hand or left-hand circular polarizer, depending on the polarity of the voltage applied to the plate. In practice an alternating voltage of 120 cycles per second is applied so that the unit alternately transmits the two circularly polarized components of the incident light. If the iron line arises in a region subjected to a magnetic field it will alternately shift in position as in Fig. 174. Now the two multipliers on the two sides of the line are connected so that their outputs are in opposition. If the intensity of sunlight is changed owing to passing cloudiness, or even if slight elliptical polarization is introduced through reflection from the mirrors of the coelostat, no change in signal

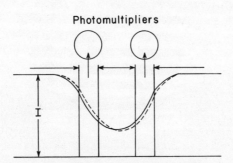

Photomultipliers

Fig. 174.   Shift in a magnetic field of the iron 5250.2-A line for right-hand (solid curve) and left-hand (dashed curve) circular polarizations.

from the two photocells is found. On the other hand, if the line shifts in position, the current in one is increased while the current in the other is decreased and a change in output results. But only if the change is due to two lines of opposite circular polarization will this output be modulated at 120 c/sec.

The amplifier following the photocells selects only the 120-c/sec component of their output and displays it on a cathode-ray oscilloscope. By driving the secondary mirror of the coelostat, motors move the image of the sun across the slit of the spectrograph in a

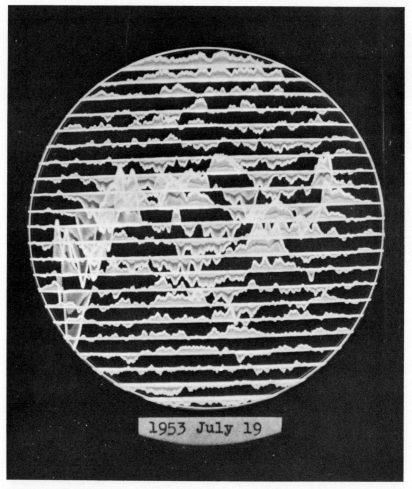

Fig. 175. Record showing magnetic fields on the solar disk. (Courtesy of Dr. Horace W. Babcock.)

pattern of about 20 lines evenly spaced on the disk from its north pole to its south pole in a period of about 50 minutes. The pattern recorded on the oscilloscope is photographed meanwhile (Fig. 175). The photograph records the line-of-sight magnetic fields encountered at every point of the solar disk. For each of the scans the deflection of the curve from the fiducial line showing the track of the scan indicates the magnitude and polarity of the field. In the figure a deflection equal to the spacing of the grid of lines results from a field of only 10 gauss. The equipment becomes overloaded near 20 gauss so that when high fields near sunspots are encountered the deflection does not go off the screen. The noise level of the equipment is only of the order of 0.1 gauss so that shifts of the 5250.2-A line of only 0.0001 of its width are detectable. The glass plate in front of the phototubes is tilted during a scan to compensate for the Doppler shift of the line due to the solar rotation. This shift approximates the line width. Since the Doppler shift is not modulated at 120 c/sec, the compensation has only to keep the line centered on the photomultipliers.

The Babcocks have found that there is a general solar magnetic field in latitudes within 20° of the poles and that it is of the order of 1 gauss. They also find that there are regions on the sun in the sunspot belts having fields of the order of 3 gauss which are not associated with any visual feature. It was very likely these fields that caused the discrepancies found by earlier investigators.

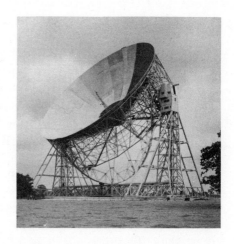

8

Radio
Telescopes

The new field of observation, radio astronomy, is now contributing greatly to knowledge of our universe. The science began in 1931 with the discovery by Karl Jansky of radio waves of extraterrestrial origin. At the Bell Laboratories he was studying the interference to radio reception that arose from static, but discovered that a large part of the radio noise was coming from the galaxy.

Jansky's important astronomical result lay dormant until Grote Reber built the first radio telescope in his back yard in 1939 and succeeded in mapping the radiation from the galaxy. But radio astronomy attracted little attention until after the war, when the military developments in sensitive receivers and narrow-beam antennas became generally available. During the war several astronomical radio discoveries were made but, owing to wartime security, they were not published until later. Two of these discoveries came about from interferences found in radar equipment which were traced to the sun. J. S. Hey found very erratic emission at several meters wavelength and traced this to sunspots. Thermal waves from the sun were looked for and measured by G. C. Southworth at 10 centimeters. Still another wartime discovery was made by Hey. He and G. S. Stewart found that meteors could be observed with radar equipment. These discoveries, augmented by the tremendously improved techniques developed during the war, gave birth to this new branch of astronomy which is now revolutionizing many of our concepts of the universe.

The science of radio astronomy was given a new dimension—spectroscopy—with the discovery by Ewen and Purcell of radio waves emitted by hydrogen in the interstellar space of our Milky Way. This discovery has made it possible to make an analysis of the structure of our galaxy. It had previously been realized that, besides the over-all emission from our galaxy, there was radiation from sources of small angular area, sometimes called radio stars. Only a few of these objects have been identified with objects seen on photographs taken with the largest reflectors. Perhaps many of these objects are peculiar galaxies far beyond the reach of our largest photographic instruments. Nearly all fields of astronomy are now benefiting by discoveries made in the radio region.

Antennas

The most important part of any telescope is that part which directs the observations to a specific part of the sky. Generally this element collects a large amount of energy as well. In ordinary telescopes the element is the telescope itself; in radio instruments it is referred to as the antenna or aerial. Directional antennas have largely been developed for communication purposes. The terminology that has resulted treats the antenna as if it were used only for transmitting. Thus the lead or cable going to the antenna is called the feed.

The directional antennas of radio telescopes take many different forms. They nearly all begin with one basic element—the dipole. This consists of two wires or rods in line with each other, supported by insulators. The ends that are adjacent are connected by a cable or other means to the input of the radio receiver. The total length of the two rods is usually slightly less than one-half wavelength of the radiation to be received, but the dipole may, according to design considerations, have other lengths. The dipole is itself a directional antenna, though one with very poor resolution. It receives best in directions perpendicular to itself and not at all in directions away from its ends.

By adding a parabolic metal reflector so that the dipole is at the focus we obtain one of the most useful antennas possible. In order to have even a moderate degree of resolution, however, the parabolic dish must have a diameter of many wavelengths. The formula, $\theta = 1.22\lambda/D$, that determines the diffraction limit to telescopic definition applies here. For a wavelength of 1 meter, an antenna with an aperture of 100 meters—truly an immense instrument—will receive from an area of the sky larger than the full moon. Radio telescopes, even for shorter wavelengths, have beam widths that are usually several degrees or larger. This points up one of the fundamental weaknesses of radio astronomy—poor spatial discrimination.

The parabolic dish is made with a very low f-number, usually about $f/0.5$. There are two reasons for such low values. A short focal length means that the dipole or other arrangement for feed of the dish does not have to be supported very far in front of it. This results in more firm support. The other reason is that it is difficult to prevent a dipole from receiving energy from most of its surroundings. If the parabola has a short focal length, the dish occupies a large part of the surroundings and shields the dipole from radiation coming from behind and illuminates the dipole with focused energy from the observed source. Frequently the dipole is further shielded by backing it with a small reflector, either a parasitic element (to be explained presently in the discussion of Yagi antennas) or a flat metal disk, so that it cannot receive energy directly from the front. The shielding of the dipole reduces interference from terrestrial radio noise and increases the efficiency of the antenna. Often a parabola is fed by a wave guide that terminates in a horn aimed at the dish. The horn prevents entry of energy coming from directions other than that of the axis of the antenna.

The choice between dipole with reflector and horn feeds depends largely on the wavelengths to be observed. For long wavelengths, the dipole is used because the horn becomes ungainly in size. The horn is considerably more efficient and so it is always used for short waves, when its size is not a handicap.

Fortunately, the dish does not have to be made of solid metal but can be in the form of mesh or wire construction. This greatly reduces the force on the antenna caused by wind and also affords a great saving in weight. The size of the holes in the mesh is determined by the shortest wavelength to be used. The mesh size should be at least as small as one-fifth of the shortest wavelength. Also, the accuracy of shape of the surface must be of the order of one-fifth wavelength, in order not to spoil the resolution of the telescope.

At least one large parabola has been made with a solid metal surface. It is the 50-foot antenna at the Naval Research Laboratory in Washington, D. C. With a surface accuracy of 1.5 mm it has been used at wavelengths as short as 8 mm.

The largest radio telescope (Fig. 176) has recently been built at Jodrell Bank in England for the University of Manchester. But its

Fig. 176.   The 250-foot radio telescope at Jodrell Bank, England. (Courtesy of Dr. A. C. B. Lovell.)

Fig. 177.   The 60-foot equatorially mounted radio telescope of the Harvard College Observatory.

250-foot diameter is less than the hypothetical one described near the beginning of this section. A telescope of 140-foot aperture is being built at Greenbank, West Virginia, and construction of one of 600-foot size is under way at Sugar Grove, West Virginia.

Other large movable parabolic antennas of this type now in use are at Dwingelo, Holland, at Bonn, Germany, and at the Naval Research Laboratory, all with apertures of 25 m. Another large radio telescope in the United States is at the George R. Agassiz Station of the Harvard Observatory and has a 60-foot aperture (Fig. 177).

We have mentioned that the directivity of the dipole antenna may be improved by the addition of other elements. These elements may be parasitic, that is, they may have no metallic connection to the receiver, being connected only by their radiative reaction with the dipole. A solid metal rod slightly longer than a half-wave dipole will act as a reflector if it is placed parallel to the dipole and about ⅛ wavelength from it. A similarly placed rod slightly shorter than one-half wavelength will act as a director; that is, it will enable the dipole to receive more energy from the sector toward the parasitic element. The Yagi or parasitic antenna is illustrated in Fig. 178. Usually only one reflector element is used but the antenna may contain three or more directors. With more than this number of elements the law of diminishing returns is reached and other methods must be used for more directive antennas.

The reflector is only about 5 percent longer and the directors 5 percent shorter than the dipole. The spacing of the elements is about ⅛ wavelength. The only practical way of determining the correct lengths and spacing is by trial and error. The beam width is of the order of 90°. Yagi antennas are familiar items as television aerials. Their chief uses in astronomy are for observations of meteors and for feeding a parabolic antenna. Obviously, they perform well only over a narrow range of wavelengths.

Another elaboration of the dipole is the use of numerous dipoles connected together. Figure 179 shows two arrangements of dipoles and their radiative or receiving patterns. In the collinear array all of the dipoles are connected so that they send the waves they col-

Fig. 178.   A Yagi directional antenna.

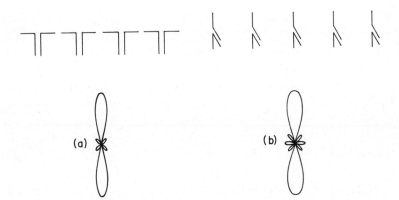

Fig. 179. Arrays of dipoles. (*a*) The collinear array receives best in directions perpendicular to the line of dipoles as in the pattern at the bottom. (*b*) The broadside array receives best in the direction perpendicular to the plane of the dipoles.

lect to the receiver in the same phase. The array receives equally well in all directions perpendicular to the line of dipoles. Accordingly the pattern shown in the figure must be rotated about a horizontal axis to obtain the three-dimensional receiving pattern. In the broadside array the dipoles are also connected so that they receive and add the incident radiations in phase. The directive pattern shown in the figure is that in the plane containing the dipoles, that is, the direction of reception is broadside to the array; the receiving pattern in a plane perpendicular to this is the same as for a single dipole. By combining the collinear and broadside arrays to form a two-dimensional array, the receiving pattern is narrowed in both the horizontal and the vertical directions, thus being restricted to a narrow beam in both directions perpendicular to the array.

Antennas of the two-dimensional type can be still further extended by adding dipoles in the third dimension. The arrangement of parallel dipoles, phased to receive most strongly in their plane instead of perpendicular to their plane as in the broadside array, is called the end-fire array. It is not found so frequently as the broadside and collinear forms, but is combined with the collinear and broadside arrays in the three-dimensional array. These arrays receive equally well from opposite directions, a distinct disadvantage. However, they do not suffer as much as the Yagi antenna from the law of diminishing returns as further elements are added. Therefore, higher directivity may be attained with arrays of dipoles.

The bidirectional property may be eliminated by the addition of a wire-mesh plane mirror behind the array of elements to form a mattress array. Figure 180 shows such as array of four elements. The distance of the dipoles from the reflector is ¼ wavelength. In practice, 64 or more such elements may be used. An array of 64 elements with reflector has a beam width of about 14°. All of these antennas are single-wavelength devices, as is the Yagi. However, the plane reflector may be used to mount several different sets of dipoles for different wavelengths without mutual interference, permitting an expensive mounting to be used for several frequencies.

Another way in which the bidirectional property can be eliminated is by forming an array from Yagi antennas instead of from dipoles. The array may be either a broadside or a collinear array or a combined array in two dimensions with the Yagis pointing in the third.

The resolution and directional pattern of all such arrays approximates to the diffraction pattern of an aperture that has the same dimensions as the extent of the antenna. For example, the 64-element antenna mentioned above was used for radar observation of the moon in 1946. The spacing between adjacent dipoles was $\lambda/2$ so that the whole array was 4λ on a side. The resolution of a square aperture of length L on a side is $57.3\lambda/L$ degrees. Replacing L by 4λ gives us an angular resolution of $14.3°$ in agreement with the experimental result for a 64-element antenna. The rec-

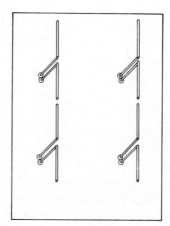

Fig. 180. The mattress array is a combination of the collinear and the broadside array with a plane reflector.

tangular aperture equivalent to an array may also be thought of as the area that is effective in receiving radio energy. Thus the larger the area the more directive and collective the antenna. The gain of the antenna is another method of expressing this. The gain is the ratio of the signal coming from the favored direction to the signal that would be received on an antenna which received equally from all directions. The ratio of the gains of two antennas is also the ratio of their equivalent areas. A single dipole, for example, has an effective area of approximately 0.1 square wavelength.

The effective area may, however, be larger than the over-all area of the antenna perpendicular to its favored direction. One example is the Yagi. A four-element Yagi has an effective area about four times that of the dipole. Another simple antenna that has a receiving area considerably larger than its apparent area is the helix. This element receives circularly polarized energy while the dipole receives linear polarizations. Its effective receiving area depends on the length and diameter of the helix but may be ten or more times the area of a dipole. A feature of the helix is that it is much less frequency-selective than the dipole, and antennas using it may receive over a wider band of wavelengths. There is also much less interaction between it and its neighbors and so the law of diminishing returns is not reached in arrays of helices. As with the dipole, the helix must be backed up by a plane reflector in order to receive predominantly from one direction. At the Ohio State University there is a large antenna using 96 helices with a plane reflector in an array 22 by 160 feet (Fig. 181). The beam of this telescope is 1° by 8° at a wavelength of 1.2 meters. It is fixed to receive only on the meridian, but the altitude may be varied so as to point to different declinations.

Antenna Mounting

The huge size required of radio telescopes in order to provide even the lowest sort of resolution, particularly at long wavelengths, places great emphasis on mounting problems. Antennas of the order of 50 m on a side are naturally unwieldy. Sometimes such antennas are fixed in position and the direction of aim is changed electronically, as will be described later in this chapter. Usually this cannot be done and the antenna is oriented in at least one coordinate.

Fig. 181.   The Ohio State University Radio Telescope consists of 96 helices mounted on a reflector 160 feet long by 22 feet wide. (Courtesy of Dr. J. D. Kraus.)

Many antennas are fixed to observe on the meridian and orientation in altitude alone is provided, as in the Ohio State telescope. The rotation of the earth on the axis sweeps the sky past the telescope's beam while the output of the receiver is amplified and recorded throughout a day. Next day, as the sky is swept through again, the altitude of the telescope is changed so that a new strip of the sky is recorded. In this way a complete map of the sky at the observed wavelength is built up. The data from the recorder are finally reduced to provide a contour map of intensity of radiation from the sky. Such a procedure is often adequate as long as the objects do not require observing for a continuous period of time, as when spectroscopy is performed or variation of an object is observed. When these are the purpose of the observer, then antennas that are steerable in two coordinates are required.

The natural mounting for telescopes is one providing equatorial coordinates—a polar axis and a declination axis. The immense size of radio antennas makes their mounting in this fashion difficult, to say the least. The weight of a large dish is enormous but in addition to its own weight the bearings and axles of the mounting must bear the weight of the counterbalances and the forces of the wind and snow load. Furthermore, the mounting should preserve accurate positions in winds up to 50 km/hr.

A large instrument that is mounted on equatorial axes is the 60-foot parabola of the Harvard Observatory (Fig. 177). The dish is firmly attached to a torque-tube declination axis. The weight of 4 tons about this axis is counterbalanced by two concrete blocks of 2 tons each. The polar axis consists of a single roller bearing, 7 feet in diameter. The concrete block at the lower right in the figure counterbalances the torque around this axis and weighs 12 tons. By selsyn-operated dials in the control house, the antenna's position can be set within 1 minute of arc and it will remain centered on this position within the same accuracy in winds up to 30 km/hr. It is also designed to withstand hurricane winds of 200 km/hr with yet a further safety factor.

It does not seem likely that the largest radio telescopes will be mounted on equatorial axes. Instead altitude-azimuth mounts will be employed. With this type of mount the problems of counterbalancing and adequate bearing support are reduced. But altitude-azimuth coordinates do not permit tracking an object by motion about one axis alone; the antenna must be moved simultaneously and precisely about both axes. This has been done with the aid of an analog computer (Fig. 182) which must be precisely made if pointing accuracy is to be maintained. A model telescope is mounted upon polar and declination axes and pointed toward the object to be examined. An additional altitude-azimuth-mounted model reproduces the position of the radio telescope through selsyn indicators. This dummy instrument is provided with switches which straddle the pointing tube of the equatorial model. If the real telescope is not oriented the same as the equatorial model, the appropriate switch closes to move the telescope in the proper direction until it is pointing correctly. This servo system has the important disadvantage that it will not function near to the zenith. Nevertheless, the largest instruments, such as the Dutch, the German, and the 250-footer at Jodrell Bank are mounted on altitude-azimuth axes and use this coordinate translator.

Fig. 182.   The analog computer for the 50-foot antenna of the Naval Research Laboratory for converting altitude-azimuth to equatorial coördinates (Courtesy of Dr. C. H. Mayer.)

Antenna Feeds

Most antennas for relatively low frequencies are fed by dipoles, and the dipole is connected to the receiver by a transmission line. Losses in this part of the telescope must be kept to a minimum; hence the first stage of the receiver should be located near the antenna to avoid long lines. Furthermore, the standing-wave ratio (the ratio of the maximum voltage along the line to the minimum voltage) must be kept as close to unity as practicable, in order to reduce losses. This type of line should also be chosen for low losses.

For the lowest frequencies the transmission line may simply be a pair of parallel wires [Fig. 183(a)] which may be kept in position by insulated spreaders or continuous plastic insulation. The impedance of the line depends on the spacing of the two wires and on their diameter. At higher frequencies, at wavelengths of a few meters, coaxial cable should be used to prevent losses from the line by radiation. This cable [Fig. 183 (b)] has a cylindrical outer conductor and a concentric inner wire. All radiofrequency currents remain inside of the hollow conductor and so radiation cannot take place. Coaxial cable with a solid dielectric is manufactured with a variety of impedances. Cable that is insulated mainly by air or nitrogen, with only occasional insulating beads, has lower attenuation.

At the shortest wavelengths, in the centimeter region, wave guides must be resorted to. The commonest form of wave guide is a tube of rectangular cross section [Fig. 183(c)], usually silver-plated on the inside. A wave can travel down the pipe by reflecting back and forth between the walls. The larger dimension of the rectangle must be at least 0.5 wavelength in order for the energy to propagate along the guide. Usually this dimension of the guide is about 0.8λ and the other side of the rectangle is about one-half this. Wave guides for the lower frequencies become quite large. For example, guides for 21-cm waves are already 18 cm wide. Wave guides find frequent use for 10-cm and shorter waves.

Usually, where wave guides are employed, parabolic antennas are used. The simplest way of feeding a dish from a guide is to terminate the guide with a flared section or horn, as shown in Fig. 183(c). The horn is aimed at the dish and its mouth is placed

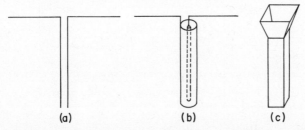

(a)   (b)   (c)

Fig. 183. Feed lines and radiators: (a) parallel transmission line connected to a dipole; (b) coaxial cable connected to a dipole; (c) waveguide terminated by a horn radiator.

at the focus. The amount of flare and the length of the tapered section are chosen so that the emerging radiation just fills the antenna. A horn and short waveguide section can be seen in Fig. 177 through the mesh of the dish.

Interferometers

During the war, bursts of noise associated with the presence of sunspots had been discovered by J. S. Hey. It was desired to determine the exact location of these noise storms on the solar disk and to compare their area with the size of the sunspot group with which they were associated. However, radio telescopes of higher resolution were needed so that the position and size of the sources might be determined. One possibility was larger antennas, but the practical limits imposed by their huge size were already being reached.

In 1946 Ryle and Vonberg solved the problem by an interferometer technique. They applied the radio analog of Young's experiment (Chapter 1). Young's interference method had already been used by A. A. Michelson in 1890 to measure the diameter of the moons of Jupiter at the Lick Observatory. Two beams of light were defined by two slits in front of the telescope objective. In 1922 Michelson and F. G. Pease further used this principle (Fig. 184) in conjunction with the 100-inch telescope to measure the diameter of one of the largest stars, Betelgeuse. Mirrors M_3 and M_4 and the two movable mirrors, M_1 and M_2, which could be separated by as much as 6 m, directed two light beams into the telescope. This

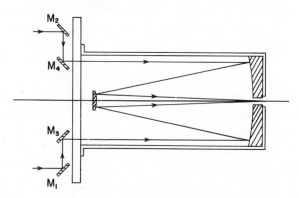

Fig. 184.   The interferometer as used by Michelson and Pease on the 100-inch telescope.

arrangement produces interference fringes of the same nature as those in Young's experiment. When the atmosphere is steady fringes may be seen with mirror separations up to 6 m. If the star is double two sets of fringes are produced, which are displaced from each other by the amount that the star images are separated. If the separation of the stars is in the same direction as the fringes and equal to one-half of the fringe spacing, the bright fringes of one star coincide with the dark fringes of the other and no fringes can be seen. In practice the movable mirrors are separated and the beam supporting them is rotated about the telescope axis until this condition is found. The star separation may then be found from the relation $\theta = \lambda/2d$, where θ is the angular distance in radians and d is the distance between mirrors M_1 and M_2. If the angle is expressed in seconds of arc, the left-hand member of the equation must be multiplied by 206,265. When a single star of sensible diameter is viewed the fringes also disappear for some separation of the two mirrors. The stellar diameter is then related to this separation by the equation $\theta = 1.22\,\lambda/d$. For Betelgeuse the fringes disappeared when the mirrors were separated by 307 cm, yielding an apparent diameter of the star of 0.047 second, which is in good agreement with the value based on its known temperature and energy output.

When the problem of insufficient resolution arose in radio astronomy, it was natural that the analog of this procedure should be tried. Figure 185 shows the radio equivalent of Young's experiment. Two directive antennas of fixed, but relatively large, separation are connected together to a single receiver. The receiving pattern in the plane containing the antennas is given at the right in the figure and consists of many lobes or regions of strong reception. The receiving pattern in the plane at right angles to the figure is merely the pattern in this plane for a single antenna. The dual system is fixed upon the earth and so sweeps the sky as the earth rotates. As a small radio source travels through the lobe pattern a record is obtained like that shown at the bottom of Fig. 185. The directivity of the individual antennas is indicated by the extent of the whole pattern, while the uncertainty in the determination of the right ascension of the sources is much less than the width of one lobe of the pattern. For very large separations of the antennas, attenuation on the transmission-line system between the antennas and the receiver becomes a problem. This difficulty can be allevi-

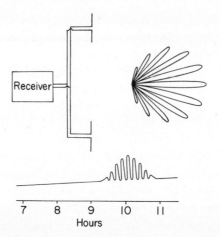

Fig. 185.   The two-antenna radio interferometer. At the upper left is the connection of the two dipoles to the receiver; at the upper right is the receiving pattern. The response produced in the receiver by a small intense source as it transits the interferometer pattern is shown a the bottom. The slope of the response curve is caused by the radiation from the Galaxy.

ated by using preamplifiers before feeding the signals to the radio lines or even by using radio links. Another difficulty that arises with the two-antenna interferometer is an uncertainty in the indentification of the central fringe. This may usually be resolved by using two different antenna spacings or even a variable spacing. The fringe that remains in the same position when referred to the sky is the central fringe and gives the position of the source.

Another type of interferometer, which needs only one antenna, is the analog of Lloyd's mirror experiment. The telescope is placed atop a cliff (Fig. 186) overlooking the sea and pointed to the horizon where the object to be studied will rise or set. The image of the antenna reflected in the sea is shown dotted in the figure. Actually the use of this interferometer in radio astronomy predates the use of the two-antenna kind. The sea interferometer was first used by Pawsey and McCready for investigating the location on the solar disk of sources of radio noise. The type of record that is obtained with this interferometer is shown in the figure. Before the source rises there is no reception of radiation; the fringe pattern begins abruptly at the time of rising. Therefore no uncertainty in the position of the source arises as with the two-antenna interferometer. Since long transmission lines are not needed, no difficulty enters here. However, there are sometimes difficulties from anomalous atmospheric refraction at the low altitudes that must be used, which may give erroneous results in positions of objects. The positions may usually be obtained in both declination and right ascension. The interferometer yields the times at which the object attains the altitudes of the different lobes of the receiving pattern. From these times both coördinates may be obtained, though usually with considerably better accuracy in one than in the other.

Sources with large area, the galaxy, for example, give a slowly varying but large signal which masks faint small-area sources. A switching technique in a two-antenna interferometer (Fig. 187) can eliminate the galactic radiation so that many more faint discrete sources can be found. By adding a phase-shifting network in the feed system of one antenna, the response pattern of the interferometer is shifted so that the major lobes fall at the positions that were previously midway between lobes. The polarity of the recording meter connected to the output of the receiver may also be changed by a switch. If the switching in the two places is performed rapidly, say 30 times per second, the meter will record the difference between the two antenna patterns. Since the broad galactic radiation is the same in the pattern with the phase shift as

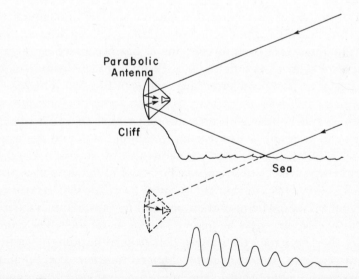

Fig. 186.   The sea interferometer mounted on a cliff is analogous to Lloyd's mirror experiment. The received signal as the source rises above the horizon is shown below.

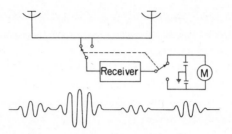

Fig. 187.   Switching technique used with the two-antenna interferometer to remove the galactic radiation and emphasize the fainter discrete sources, as in the record at the bottom.

without it, it will cancel out. Only the radio stars will be recorded, with oscillatory records as shown in the figure. Many discrete radio sources have been found by this technique.

The main difficulty of the two-beam interferometers is the ambiguity that is usually present in interpreting the observed patterns when two or more sources are within the beam. The difficulty exists most of the time when the last-described system, with its increased sensitivity to weak sources, is employed. Not only are the observed patterns complex but there is often an unresolvable multiplicity of the arrangement of sources that will satisfy the observed patterns.

A considerable improvement in eliminating the ambiguities came about through the introduction of interferometers with more than two antennas. One such interferometer has been used on 20 centimeters for solar observations. Here 16 small parabolic antennas all lying in a line were connected together as in Fig. 188. The response pattern of such an arrangement, is also shown in the diagram, produces several knife-edge beams through which an observed object sweeps as a result of the earth's rotation. At lower frequencies simple dipoles may be used, or elementary directive arrays. The resulting received pattern approaches those of the broadside arrays or collinear arrays in that the pattern has only one or at most a few major lobes. The resolution, depending on the linear extent of the interferometer, may be quite high—perhaps only a few minutes of arc. Not only may the pattern be left fixed in position while the sky sweeps past; it may even be changed in orientation by adding a progressive phase shift in the lines from individual antennas so that the phase of each antenna in the line is shifted with respect to that of the previous one. By continuously increasing the amount of the progressive shift the lobe pattern may be swung to follow an object.

All of the interferometers that have been described so far have high resolution in a plane containing the line of the elements but usually poor resolution in the plane perpendicular to this. Two interferometers may be built at right angles to each other and the patterns received separately on the two. However, an ambiguity still results in interpreting a complicated array of sources. What must be done is to combine the two interferometers so that only a unique pencil beam results where single major lobes of the two patterns cross. B. Y. Mills in Australia first combined two interferometers to achieve this effect in an instrument now known as the Mills cross. A photograph of this instrument is shown in Fig. 189. Each arm is 500 yards long and contains 500 half-wave dipoles above a wire-mesh reflector. The beam width is 0.8° at a wavelength of 3.5 meters.

The method by which the two arms are combined to give a single pencil beam is described in Fig. 190. At the left of the figure is the scheme of the interferometer with arms A_1 and A_2. In the center of the drawing, the response a_1 of the principal lobe is shown; the response a_2 of the other arm is similar but at right angles to a_1. If now the two interferometers are connected together so that their outputs add before they are connected to the receiver, the pattern will be as shown in (b). A source occurring at the intersection of the two lobes will be strongly received, a source occurring in only one of the lobes will be received at only half of this intensity, and a source in neither of the lobes will be received very weakly. Now if the two arms are connected so that they subtract from one another, as can be done by shifting the phase of one interferometer by 180° before connecting to the other and to the receiver, a reception pattern as in (c) will be obtained. A source

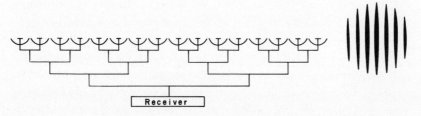

Fig. 188.   The Christianson interferometer and its receiving pattern.

Fig. 189.   The original Mills cross antenna. Each arm is 500 yards long. (Courtesy of *Sky and Telescope*.)

at the intersection of the two major lobes will give equal signals in both arms and so will cancel out in the subtraction. A source in only one of the lobes will remain and a source that is in neither will be only weakly received as before. A switch is arranged to alternate rapidly between these two connections while synchronously with the alternation the connections to the recording meter are switched as was described previously in the beam-switching interferometer. The parts of the two patterns that are received equally will now cancel in the recording meter and only for a source in the center beam at the intersection of the lobes of the two arms is there an appreciable difference. The result is a pencil receiving beam.

This beam, if all of the dipoles in each arm are added in phase, will be vertically upward. However, it may be moved about by adding progressive phase increments between successive dipoles in an arm. Customarily the cross is laid out in north-south and east-west directions. If the phase increments are made in the north-south arm, the declination of the beam is altered. Frequently, changes are made only in this arm so that the telescope is a meridian transit instrument and the terrestrial rotation is depended upon for sweeping the sky past it.

The primary feature of the cross is its high resolution; it is equivalent to that of a parabolic dish having the extent of the arms. However, an important point must be made. The Mills cross does not in the least eliminate the necessity for large parabolic antennas. The cross has a number of disadvantages. It is nearly a one-wavelength device because to change to another frequency the lengths of all of the dipoles have to be changed. Further, the cross does not have the energy-collecting ability of a dish of equivalent size. At low frequencies, where the cross is customarily employed, this is not a serious disadvantage, but in the region of the 21-cm hydrogen emission its small effective area is a definite disadvantage. The pointing of the interferometer cannot be readily changed to track an object in its east-west motion and so spectroscopy and other types of programs requiring continuous observing are not feasible. What the Mills cross does yield is a very narrow beam for observations on wavelengths in the meter range where the size of equally directive parabolas would be impossibly large. An interferometer cross was recently constructed by the Department of Terrestrial Magnetism of the Carnegie Institution. It has arms of 1500 meters length and operates on a frequency of 22 megacycles per second, or about 14 meters wavelength.

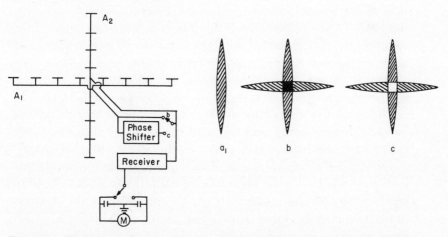

Fig. 190.   The connections of the two arms of the Mills cross and the receiving pattern of (*a*) one arm, (*b*) two arms connected in phase, and (*c*) two arms connected out of phase.

Radio Receivers and Radiometers

An important part of the radio telescope is the receiver. It is very much like an ordinary radio receiver in its electrical circuitry but it is much more carefully built, for the sensitivity of the telescope depends to a large extent on the receiver. When the receiver is connected to a recording meter and the system is calibrated by some means to read power incident on the antenna, the whole system becomes a radiometer.

The function of the receiver is to amplify the incoming signals to appreciable levels, to select a specific band of frequencies of predetermined width and frequency, and to convert the amplified signal from one that alternates at the signal frequency to one that varies proportionally only to the amplitude of the original signal. The last process, that of conversion, is usually called detection. All of these steps the receiver must perform with the introduction of a minimum amount of noise of its own. The circuit that is customarily utilized for this purpose is the superheterodyne circuit invented by E. H. Armstrong in 1919. The functioning of the circuit and the associated changes occurring in the signal at each point are depicted by the block diagram of Fig. 191. The signal from the antenna is first amplified by a radio-frequency amplifier, though often this stage is omitted. The signal is then inserted into a nonlinear circuit, which may be a properly biased vacuum tube or a silicon crystal, along with a signal from the local oscillator. Because of the nonlinearity of the mixer, frequencies equal to the sum and the difference between the signal and local-oscillator frequencies appear in the mixer output. If the local-oscillator frequency is chosen near to the signal the difference will be a low frequency, perhaps 30 or even as low as 0.5 megacycle/second. The intermediate-frequency amplifier is tuned to select a sharply defined band at this frequency and reject all others. Most of the receiver's gain or amplification takes place in the intermediate-frequency amplifier, for at 30 Mc/sec and below amplifiers are stable and introduce relatively little noise of their own. If a wide band is to be received, say several megacycles per second or more, the intermediate frequency is chosen rather high, often 30 Mc/sec; if rather sharp tuning is required, low frequencies are used. The desired part of the original signal is customarily its amplitude or perhaps the

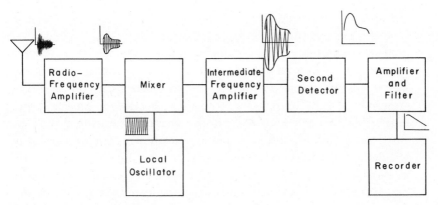

Fig. 191. Block diagram of a superhetrodyne receiver. The insets show the nature of the signal at each stage.

variation of its amplitude in time. To obtain just this part, the intermediate frequency is rectified in the second detector (the mixer is sometimes referred to as the first detector). The output is further amplified and electrically filtered so that its band width lies between zero and some frequency of the order of several thousand cycles per second, or perhaps only 1 c/sec or less. Finally, the amplified and filtered signal is connected to a pen-and-ink recorder.

The frequency that is received may be changed by altering the frequency of the local oscillator. Sometimes the change is by a factor of as much as 2; usually, however, circuits are not designed for such wide coverage.

A difficulty that is encountered in superheterodyne circuits is the reception of another frequency, the image, for there are two frequencies whose difference from that of the local oscillator is just the intermediate frequency—one above the local-oscillator frequency and one below. The undesired one of these two frequencies is rejected by tuned circuits at the input of the receiver and in the radio-frequency amplifier, but the degree of rejection at high frequencies may be poor. The rejection may be made better, the higher is the intermediate frequency, and so high-frequency intermediate amplifiers are sometimes used to provide the rejection. If the sharp selectivity of a low intermediate frequency is also desired, a double superheterodyne circuit may be employed. Here, after amplification and rejection of the image at about 30 Mc/sec or higher, a second local oscillator and mixer are employed to further reduce the signal frequency to perhaps 0.5 Mc/sec where further intermediate-frequency amplification and selection occur before final rectification.

The gain offered by a receiver does not determine its sensitivity. The important characteristic is the ratio of the signal power to the noise power introduced by the receiver. This ratio is usually determined by the early stages of amplification. A resistor at a given temperature or an antenna placed inside an opaque box at a specific temperature generates an amount of noise governed only by physical laws. When either the resistor or the antenna is connected to a receiver, the performance of the receiver may be judged by the total noise power obtained at the receiver output relative to the power that would be obtained if the antenna or resistor alone were responsible for the noise. The ratio of total noise power to resistor noise power when the resistor temperature is 300° absolute is called the noise factor of the receiver. The lower the noise factor, the more sensitive is the receiver.

At wavelengths of 1 m and longer the factor may reduce to 2 or even less. At 21 cm noise factors of the order of 10 are customarily attained, while at 1 cm they may be as high as 50. Thus at 21 cm the noise generated within the receiver is equivalent to that obtained from an antenna contained in a box at a temperature of 3000° absolute. If the antenna is pointed instead at a black body at 3000°, the receiver signal will double as a result of the signal from the black body. If on the other hand the antenna is aimed at a body at a temperature of only 3° absolute the receiver output will be increased by 0.001 of its normal output. In order to measure temperatures this low (achievable in practice) the receiver gain must be stabilized to 0.1 percent, except in cases where special circuits are used, as in the Dicke radiometer.

A receiver that is stabilized and that may be accurately calibrated, so that, from the measured output from the receiver, the noise power at the antenna terminals may be determined, is called a radiometer. The equipment used for calibrating receivers will be described in the next section. It is important to note that as radio astronomy progresses absolute calibration of receivers by internal equipment will be required less and less. So long as a receiver will maintain the same amplification for long periods of time, the entire radio telescope may be occasionally calibrated by observation of standard celestial radio sources. Thus the same procedure that is applied in photoelectric photometry is taken over into radio astronomy. However, stabilization of the receiver is an all-important requirement of a radiometer and internal means should be provided for testing the constancy of receiver gain.

We have already mentioned that the gain, by careful design and use of regulated voltage supplies, may be held constant to one part in a thousand. Stability may be obtained in another manner, as was done by Dicke in his radiometer (Fig. 192). The input to the receiver is switched rapidly (at about 30 times per second) between the antenna and a resistor whose temperature is accurately controlled. The output of the receiver is simultaneously switched so that it is alternately connected to the recording meter with one and then with the opposite polarity. A smoothing circuit is employed before the meter to permit only the direct-current component to reach the meter. By this technique only the difference in the noise voltages from the antenna and the resistor is recorded on the meter, that is, the large signal due to the receiver noise is canceled. As a result a high gain stability is not required in the receiver. A low noise factor for the receiver is still desired, however, to reduce output fluctuations. The Dicke system, because it is connected to the antenna for only one-half of the time, loses a factor of 2 in sensitivity. Dicke has given an expression for the smallest temperature difference that his radiometer will measure. The formula shows that it is desirable to have a large band width in the intermediate frequency and a large time constant in the output, as well as a low noise factor in the front end. These considerations apply as well to the gain-stabilized receiver. The large time constant and band width sometimes conflict with other considerations, such as the necessity of a narrow band width in a radio spectrograph.

A way of amplifying microwaves that has extremely small noise factor (near one) has recently been discovered. The device is called the *maser,* an abbreviation for *m*icrowave *a*mplification by *s*timu-

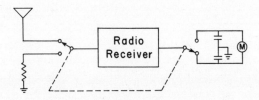

Fig. 192.   The Dicke radiometer circuit.

lated *emission* of *r*adiation. The first successful maser was operated
at Bell Telephone Laboratories in 1956.

Having no tubes and their noise-producing electrons, the device
uses a crystal containing magnetic atoms at a temperature within
a few degrees of absolute zero. In Chapter 1 we mentioned that
gases often emit and absorb light in lines in the visible part of the
spectrum. A similar phenomenon occurs in the emission or absorption of microwaves by certain solids when they are subjected to a
magnetic field. The emission or absorption occurs when an atom
composing the solid changes its state of energy. Figure 193 shows
three possible energy states of chromium atoms in a ruby crystal.
There are always many atoms occupying each of the three states.
A change in energy of an atom by its moving from state 3 to state
1 is accompanied by emission of microwave energy of frequency
$\nu_{1,3}$. Conversely a change from state 1 to state 3 results from absorption of a photon with frequency $\nu_{1,3}$.

Of fundamental importance to the operation of the maser is the
fact that radiation of frequency $\nu_{1,3}$ may induce an atom in state 3
to change to state 1 with emission of a photon of frequency $\nu_{1,3}$.
Whether the change from 1 to 3 and absorption of radiation or the
change from 3 to 1 and resultant emission of radiation is more frequent depends on the relative number of atoms in the two states.
If there are more atoms in state 1 than in state 3, then net absorption of radiation results when microwave energy of frequency $\nu_{1,3}$
passes through the crystal. Net emission occurs if more atoms are
in state 3. If this condition prevails, a wave of frequency $\nu_{1,3}$ passing through the crystal emerges stronger than when it entered as
a result of the contribution from atoms changing from state 3 to 1.
Amplification of the wave has taken place.

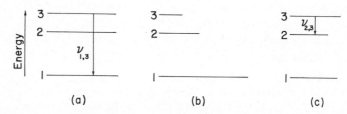

Fig. 193.   Three energy states of a maser crystal. The lengths of the lines are proportional to the number of atoms in each energy state. In (*a*) the crystal is at room
temperature; in (*b*) the crystal is cooled to a low temperature; in (*c*) radiation of
frequency $\nu_{1,3}$ is applied to the cooled crystal.

At room temperature the relative numbers of atoms in states 1, 2, and 3 are nearly equal, as shown by the lengths of the lines in Fig. 193(a). When the crystal is cooled by liquid helium to $4°$K the situation is as shown in (b); most of the atoms are in state 1. If now sufficient microwave power of frequency $\nu_{1,3}$ (only a few hundreths of a watt are required) is passed through the crystal, atoms will be moved from 1 to 3 until the numbers in 3 and 1 are nearly the same, as shown in (c); state 2 remains unchanged. With still more radiation as many atoms are moved from 3 to 1 as from 1 to 3. But we now look at levels 3 and 2; they fulfill the requirement for amplifying radiation of frequency $\nu_{2,3}$ in that the higher energy state 3 has more atoms in it than state 2 does. A wave of frequency $\nu_{2,3}$ will induce atoms in state 3 to change to 2 with emission of photons and consequent amplification of the wave.

Why is such a low temperature used? The first reason is only for producing the conditions as in (c), but there are other methods for doing this. The second reason is much more fundamental. At room temperature atoms spontaneously fall from state 3 to 1 or 2 with emission of $\nu_{1,3}$ or $\nu_{2,3}$. Such spontaneous emission does not constitute amplification because it occurs whether $\nu_{1,3}$ or $\nu_{2,3}$ is present or not. This emission does constitute noise, however. Only lowering the temperature lessens this source of noise. A maser at $4°$K is about ½₀ as noisy as one at $77°$K, the temperature of liquid nitrogen.

So far we have only discussed how amplification at frequency $\nu_{2,3}$ is produced. Suppose it is desired to amplify another frequency. The actual value of $\nu_{2,3}$ depends not only on the properties of the crystal, but also on the strength of the magnetic field to which it is subjected. The amplifying frequency is tuned, therefore, by changing the magnetic field.

The maser is only an amplifier; it must be followed by a radio receiver. It may serve as the radio-frequency amplifier in Fig. 191 and amplify the signal from the antenna to the point where noise from tubes in the remaining part of the receiver is much smaller in comparison to the signal.

Using liquid helium in a telescope involves considerable difficulties. That these difficulties are not insurmountable is proved by the use of a maser on the 50-foot telescope of the Naval Research Laboratory in 1958 to measure the temperatures of planets, and more recently elsewhere.

Calibration Sources

One way in which the receiver can be calibrated is through the use of a signal generator that provides a known output. Such a device provides a sine-wave signal which is very different from the random noise characteristic of astronomical signals. In order to provide a quantitative calibration for comparison with the wideband celestial signals, the generator frequency must be swept through the receiver pass band.

It is much more convenient and probably more accurate if the same type of signal, a noise signal, is used for calibration. Several different types of generator are available for producing known amounts of noise signal. One of the simplest of these is just a resistor at an adjustable but known temperature, or alternatively two resistors at different temperatures. The noise output of a resistor is proportional to its temperature. The resistance should be the same as that of the antenna. Such a resistor noise source may take several forms, depending on the frequency and the temperature at which it is operated. At wavelengths of a few meters or longer a lamp filament may be used. Temperatures up to 2000° are easily obtainable with a lamp, but the resistance depends on the temperature. In place of a resistor a piece of transmission line or coaxial cable or wave guide that is especially constructed to have high attenuation is placed in an oven and connected to the receiver. Hot loads of this type cannot be made for very high temperatures and consequently large outputs. For low frequencies a noise diode may be used. This is a vacuum tube with an anode and tungsten filament. The number of electrons that the tungsten filament will emit depends on its temperature. The electrons are emitted at random times and so the current to the anode fluctuates, the amount of noise depending in a known way on the current. This type of noise generator produces large signals but cannot usually be used on frequencies above 200 Mc/sec.

At frequencies where wave guides are convenient the gas discharge lamp finds use. This is not an absolute calibration source but it is an excellent secondary standard because, although its output cannot be calculated from theory, it is constant, largely independent of operating conditions. In physical form it customarily is a fluorescent lamp inserted diagonally along a wave guide at about a 10° angle. Once calibrated, it can be used from day to day to check constancy of the gain of the radiometer. Its effective temperature is between 10,000° and 20,000° absolute. Consequently it provides a signal that is as large as the noise inherent in most receivers.

Radio Spectrographs

A most useful tool for galactic research appeared in 1951 with the development of a spectrometer for a radio telescope by Ewen and Purcell and the ensuing detection of the 21-cm radiation from neutral hydrogen.

The equipmental requirements for detecting this radiation are very severe. The line width is only a few tens of kilocycles and receiver band widths as small as 5 kc/sec are frequently desired. This is in conflict with the Dicke formula, which requires large band widths for high sensitivity. The effective temperature of the radiation is 100° absolute or less, while the noise generated within the receiver is equivalent to 3000°. It is desirable to be able to measure the 100° temperature to an accuracy of a few percent, so that a sensitivity to antenna temperature changes of the order of 3° is required. Therefore the receiver output must be measurable to 0.1 percent. Highly stable equipment is necessary.

A radio spectrograph is made from a receiver in several ways. The local-oscillator frequency may be swept through the range required to make the receiver sweep the desired frequency band. Another way of accomplishing the same purpose is to provide a wide-band intermediate frequency and sweep the frequency of a sharply tuned second detector. A recorder connected to the receiver output records the energy variations as the frequency varies. Here again high stability is required, for with band widths of 5 kc/sec at 1420 Mc/sec the frequency should be known at any time during the sweep to a few parts in 10 million. Furthermore, the receiver must have uniform sensitivity over the scanned spectral range, which for 21-cm radiation is 2 Mc/sec.

The gain stability is obtained by using a modified Dicke radiometer circuit. In the normal Dicke scheme the signal from the antenna is compared with that obtained from a resistor. In the hydrogen-line receiver the signal received on the signal band is compared with the signal obtained on another band of frequencies which is just removed from the hydrogen-line frequencies. In this way, most of the effects of interfering signals and gain instabilities of the spectrograph are eliminated. The comparison frequency band is about 3 Mc/sec from the instantaneous receiver frequency. As the receiver frequency changes, the frequency of the comparison band sweeps, too. The comparison-band width is customarily large to reduce noise effects arising here.

A block diagram of a spectrograph that has been in use at the Harvard Observatory since 1953 is shown in Fig. 194. The local oscillator is tuned by a servo mechanism and "locks in" to give a beat at a frequency between 5 and 7 Mc/sec with the 280th harmonic of a 5-Mc/sec oscillator. The oscillator is synchronized to within a few cycles per second with the precise frequency broadcast by WWV, the National Bureau of Standards radio station, and maintains adequate frequency stability. A frequency-modulation receiver is tuned to the 5-to-7-Mc/sec difference frequency and its output controls the frequency of the main local oscillator. The frequency-modulation receiver is tuned through the 2-Mc/sec range by a scanning motor and at the same time the servo motor tunes the local oscillator so as to keep the difference between it and

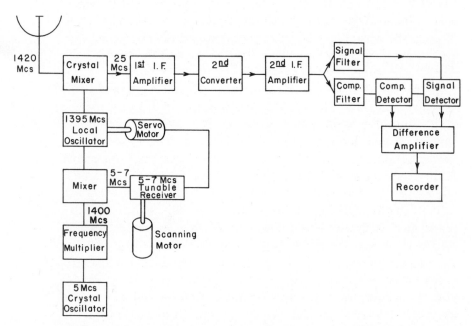

Fig. 194.   Block diagram of a radio spectrograph for the hydrogen line at 1420 Mc/s.

the 1400 Mc/sec from the 5-Mc/sec crystal equal to the frequency-modulation receiver frequency. As the receiver scans through the 2-Mc/sec range, the local oscillator follows and along with it the tuning of the 21-cm spectrograph.

From the diagram it can be seen that the main receiver is a double superheterodyne circuit. The second converter also uses an oscillator frequency derived from the highly stable 5-Mc/sec oscillator. The output of the second intermediate-frequency amplifier is split into two bands, one for the hydrogen signal and the other for the comparison band. The band width of the hydrogen signal is selected by a switch from the values 5, 15, 50, or 200 kc/sec, while the comparison-band width is always 2 Mc/sec. The output of the comparison band is reduced to make it comparable to the 21-cm signal. The comparison-band output is then subtracted from the signal output and the difference signal is applied to the recorder. If the gain of the receiver changes slightly both of these bands will vary similarly, but their difference, which is only a small fraction of either signal, will vary only slightly.

As the spectrum is scanned the frequency must be accurately recorded on the chart. This is done by a marker system which derives its accuracy from the stable oscillator. At every 5 kc/sec through the scan a special pen puts ink marks on the record. The time required to make one scan is long. If the 5-kc/sec band width is used, one spectrum is made in 4 hours.

Another field in which radio spectroscopy has been employed is solar research. Here the problem is considerably different, for the desired spectral band is several hundred megacycles per second wide at frequencies of only a few hundred megacycles per second. Moreover, this band must be swept through several times a second. Fortunately, the sun is a strong source and long signal averaging is not required as it is for detection of hydrogen radiation.

The Harvard Observatory has established such a radio spectrograph at Fort Davis, Texas. The antenna is a 28-foot paraboloid which simultaneously feeds three separate receivers. These cover the ranges 90–180, 170–320, and 300–580 Mc/sec. These frequency ranges are swept through as rapidly as three times per second. The outputs of the three receivers are displayed on cathode-ray oscilloscopes where they may be photographed on motion-picture film. The spectral distribution of solar radio bursts may change very rapidly, as indicated by Fig. 195, which shows one of the spectra obtained with this equipment. In this photograph the frequency is displayed vertically, with low frequencies at the top, and time increases to the right; the intensity of the solar energy is indicated by the brightness of the diagram. Characteristically one finds that the frequency of a burst decreases as time progresses. Much is being learned about the solar atmosphere through such radio studies.

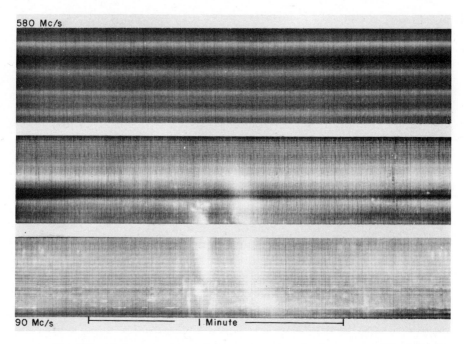

Fig. 195.   A frequency scan of the sun covering the three ranges 90–180 Mc/s, 170–320 Mc/s, and 300–580 Mc/s.

Index